D1544645

THE GREAT
BOOK OF
VEGETABLES

Antonella Palazzi

THE GREAT
BOOK OF
VEGETABLES

SIMON & SCHUSTER

LONDON·SYDNEY·NEW YORK·TOKYO·SINGAPORE·TORONTO

Photographs
Jacket: F. Pizzochero
ADNA: 294–295, 297 (A. Brusaferri)
Mondadori Archives: 41 (ADNA); 48, 159, 208
    (R. Marcialis);
42 (Visual Food); 103, 104, 117, 160, 207
L. Cretti: 99c, 100c
M. Lodi: 248, 266–267, 269, 270–271, 273, 274–275, 277,
    278–279, 281, 282–283, 285, 286–287, 289, 290–291, 293,
    298–299, 301, 302–303, 305, 306–307, 309, 310–311, 313
    F. Marcarini: 1, 2, 10, 11, 12, 13, 14, 15, 16, 17, 18, 19,
    20, 21, 22, 23, 24, 25, 26, 27, 28, 29, 30, 31, 32, 33, 34, 43,
    44, 45, 46, 99, 100, 101, 102, 155, 156, 157, 158, 203, 204,
    205, 206, 243, 244, 245, 246
R. Marcialis: 45ar, 46ar, 99bl, 99c, 100ar
G. Pisacane: 153, 201, 202, 202, 207, 241, 242, 247
F. Pizzochero: 97
Home economist
A. Avallone: (jacket) and pp. 294–295, 297
C. Dumas: 248, 266–267, 269, 270–271, 273, 274–275, 277,
    278–279, 281, 282–283, 285, 286–287, 289, 290–291, 293,
    298–299, 301, 302–303, 305, 306–307, 309, 310–311, 313

Translated by Sara Harris
Index compiled by Janey Pyke

Copyright © 1991 Arnoldo Mondadori Editore
S.p.A., Milan
English translation copyright © 1992 Arnoldo
Mondadori Editore S.p.A., Milan

This edition published 1993 by
Simon & Schuster Ltd
West Garden Place
Kendal Street
London W2 2AQ

A Paramount Communications Company

A CIP catalogue record for this book
is available from the British Library

ISBN 0-671-71162-8

Typeset by Tradespools Ltd, Frome, Somerset
Printed and bound in Italy by Arnoldo Mondadori
Editore, Verona, Italy

·Contents·

provide an almost limitless variety of taste and texture in our everyday food.

Vegetables are usually classified according to the botanical families to which they belong: leguminosae, cucurbitaceae, solanaceae, etc. but for the purposes of a cook book it is more practical to group them according to the parts we eat: leaf and stalk vegetables; shoots and fruits; bulbs, roots, tubers and seeds. A separate section is devoted to those rather special products of the soil: fungi or mushrooms and truffles.

Origins and distribution of vegetables

We know that cereals and vegetables have grown on earth for many millions of years. Man, or his anthropoid ape ancestors, relied on them for nourishment at a very early evolutionary stage, gathering and eating these plants wherever they grew, in their natural habitats. These wild varieties were very different, in both colour and shape, to those we know today. As man developed, he became more selective and learned how to cultivate certain particularly valuable plants to help him survive. Vegetables were chosen for their nutritional value and for their taste. The earliest known cultivated vegetable is the pea, which was grown in Turkey as far back as the sixth century B.C. Three centuries later, in fertile Mesopotamia, beetroot, onions, garlic and leeks were being cultivated. The first large-scale, methodical vegetable growers lived in the Middle and Far East. Turnips were prized by the inhabitants of these regions for their high starch content and were first grown by the Chinese, as were cucumbers, radishes and cabbages; these last were preserved by means of pickling, to be eaten with rice as the staple diet of the poor. Soon vegetable growing spread throughout the Mediterranean basin. The occupying Roman armies introduced leeks, cauliflowers, artichokes and carrots to Britain. The carrots of ancient times were red, violet or black, rich in

antocianine but devoid of carotene, nothing like those we eat today which were developed in Holland hundreds of years later. In the seventh century Moorish invaders conquered much of Spain and introduced aubergines and spinach; soon these were being cultivated in southern France as well. During the Middle Ages vegetables were grown extensively throughout Europe. The Dutch were particularly successful market gardeners, thanks to their fertile alluvial soil.

In the fifteenth century, when the Spanish conquistadores first started to explore North America, they discovered vegetables they had never seen before: corn, tomatoes, sweet peppers, pumpkins and the staple food of the Incas, the potato, grown successfully at altitudes above 14,000 ft (4,500 m). A two-way traffic of vegetable plants began from the New World to the Old and vice versa, with crazes or fashions for a particular, new vegetable running their course and then falling out of favour. The potato was enthusiastically adopted in England and Ireland from the seventeenth century onwards but was banned in France for a long time because it was suspected of carrying leprosy. Eventually, however, each and every one of the new American vegetables found its place in the cookery of various European cuisines.

The Frenchman Nicholas Appert discovered a safe canning method in the late eighteenth century and was able to put his invention into practice in the early nineteenth century; this was the first step towards large scale distribution and sale of vegetables out of their natural seasons. When deep-freezing was developed in the United States in the 1930s, it was only a matter of time before virtually every type of vegetable was available all the year round.

Composition and nutritional value

Most vegetables have a very high water content, a low calorific value and contain many of the substances vital to man's health: mineral salts, vitamins, organic acids, carbohydrates, fibre and very low quantities of fat.

The proportion of these elements varies from one vegetable to another: potatoes are very rich in starch, marrow has a very high water content, while carrots, onions and corn contain a lot of saccharose. Vegetable plants whose 'fruits' we eat are usually rich in potassium; leaf and 'flower' vegetables are full of iron.

With the exception of legumes, which have a high protein content, vegetables have only traces of protein, varying from 1 per cent to 4 per cent. Vegans (strict vegetarians who eat no animal products at all, including dairy produce) have to ensure that they eat plenty of legumes (beans, peas, lentils) preferably at the same time as cereals which maximize the absorption of protein.

Not all vitamins are present in vegetables: vitamins A and C predominate and, together with mineral salts, are found mainly in the larger, older, external leaves which are all too often discarded. A proportion of the vitamin content is also liable to be lost if the vegetable is not used fairly soon after it is harvested. Prolonged soaking in water and certain cooking methods further deplete their nutritional value.

All vegetables contain organic acids which are of crucial importance in maintaining the alkaline balance in our bodies and in helping to eliminate toxic substances; they neutralize the acids which we produce as a result of physical or psychological stress and help to counteract the harmful side effects of a diet too high in proteins and animal fats.

Fibre, in the form of cellulose and lignin, simply works its way down the digestive tract and is not absorbed into the bloodstream. This does not mean we can do without these substances: by their mere presence they help the body to eliminate potentially carcinogenic elements present in certain foods; they also help to lower the level of glucose and fatty acids in the blood and prevent chronic constipation and its consequence, painful diverticulitis, which frequently accompany an over-refined diet.

Legumes have special properties: unlike other vegetables they have a high energy value being rich in carbohydrate and protein. This is vegetable protein, often referred to as 'noble protein', which is much better suited to man's digestive system than animal protein. Vegetables may be called 'the poor man's meat' because they are so cheap and plentiful but an increasing number of people who can well afford to pay for the most expensive cuts of meat are opting for this healthy alternative for some, if not all, of their meals.

Note

All recipes serve 4, except where otherwise indicated.

Measurements are given in Imperial and metric. Follow one or other system, but do not combine the two.

·PREPARATION·

A few special utensils make vegetable preparation easier. Potato peelers with (1) fixed blade and (2) swivel blade; a three-pronged fork (3); a potato ricer or masher (4) and the traditional vegetable mill (5); a mandoline or vegetable slicer with adjustable blade (6). A chopping board is indispensable, a half-moon cutter (7), also known as an hachoir *or* mezzaluna *is extremely useful. Several very sharp knives are essential: a straight-bladed peeling knife (8); a curved-blade peeling knife (9); a knife with a serrated edge (10); a straight-bladed vegetable knife (11); a medium-sized filleting knife (12); a large kitchen knife (13). Scoops or melon ballers (14); a pair of good kitchen scissors (15); a brush, useful for cleaning mushrooms (16) and, very much an optional extra, a truffle slicer (17). A fluter (18), a canelle knife (19) and a zester (20). Three bowls (21) will be useful: one to hold the unprepared produce, one for peelings and trimmings and one for the prepared vegetables. A sieve (22) and a colander (23) complete the vegetable cook's basic equipment.*

☐ ONION

Use a small peeling knife with a very sharp curved blade or serrated blade to cut off the root end of the onion. Insert the point of the knife ¼–½ (1 cm) deep (1) and cut out a small cone-shaped section by turning the onion round (2) to get rid of the tough, woody section; if it is to be cooked whole, do not cut out this cone-shaped section but simply make two intersecting cuts across the base: the onion will cook evenly but will not fall apart. Peel the onion (3), removing the skin neatly. To slice: cut lengthways in half, place cut side down on the chopping board and cut lengthwise into thin slices, holding it with your fingers away from the blade. To chop the onion, hold the long slices together and cut at right angles across them, keeping your fingertips away from the blade (4) or folded under as you cut further into the onion to avoid cutting yourself.

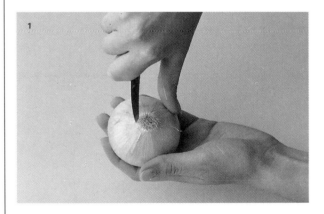

SPRING ONION, LEEK

Cut off the roots, remove the outer layer and the tough, ragged ends of the green leaves (1).

Except when preparing spring onions for crudités and salads, use the white bulb only and slice it thinly into rings. Take great care to wash leeks very thoroughly; they are often grown in sandy soil which lodges between their leaves, frequently far down towards the root. Slit the leek lengthwise (2) from top to middle and bottom to middle and then rinse between the layers under running cold water.

CHIVES

Trim a very small amount off both ends of the chives. Rinse briefly under running cold water, drain and dry with kitchen paper. Gather into a bunch, making sure all the ends are level and snip very short lengths into a bowl with kitchen scissors.

SHALLOT

Use a small peeling knife to cut off the roots and carefully pull off the outer skin, working from the roots towards the top.

To chop, use the same method as described for the onion.

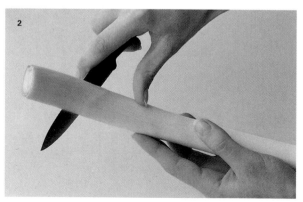

☐ GARLIC

To separate the cloves, place the flat of the blade of a large kitchen knife on top of the head of garlic and bring your fist sharply down on the flat surface of the blade (1). Use a small peeling knife with a curved or straight blade to peel the cloves one by one, pulling the skin away from the base to the top (2). Crush the cloves if necessary by placing the large knife blade flat on top of the garlic clove and give it the same, sharp blow with your fist as before.

☐ POTATO, CELERIAC, TURNIP

Peel with a small, sharp peeling knife with a curved or straight blade or with your usual potato peeler. Use the point of the knife or peeler to take out every trace of black spots or eyes (1). To avoid burning your hands when peeling very hot potatoes that have been boiled in their skins, spear the potato with a special three-pronged fork if you have one (2), or with a carving fork to hold the potato steady while you remove its skin.

☐ TOMATOES

Remove the stem with the tip of a sharp knife (1). The easiest way to skin tomatoes is to blanch them first: make a cross-shaped incision in the tomato (2) then plunge in boiling water for 10–20 seconds. Drain and refresh under running cold water to arrest the cooking process. Drain again and peel off the skin with the tip of a knife. It will come away easily. Quarter the tomatoes and extract the seeds and pulp (3). To prepare tomatoes for stuffing, slice horizontally in half (4) then extract all the pulp. Drain well and spoon in your chosen stuffing.

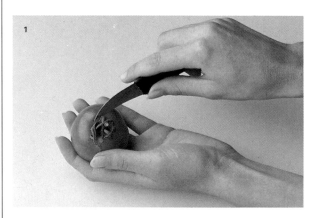

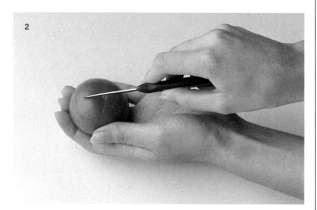

CARROT, CUCUMBER, DAIKON ROOT, SCORZONERA, PARSNIP

All these vegetables can be prepared in the same way. Cut off the stalk end (1) and the tip, then hold the vegetable firmly in one hand while running the potato peeler from the larger, stalk end to the tip (2) to remove a thin, outer layer all the way round.

When preparing scorzonera (also called black salsify), have a bowl of cold water mixed with lemon juice standing ready and drop the pieces of vegetable into it as soon as you have peeled them to prevent discoloration which starts as soon as the outer, protective layer is removed and the flesh is exposed to the air.

AUBERGINE

Slice off the stalk and the tip, removing a generous slice to get rid of the tough flesh and bitter skin at each end. The rest of the skin does not have to be removed, so peeling is optional. If you intend to fry the aubergine, you can cut it lengthwise in half and then slice across as thickly or thinly as required. When small cubes or dice or rectangles are more suitable (for casseroles etc.), cut lengthwise into quarters (1) cut out the central seed-bearing section (2). Cut each quarter lengthwise in half and then slice across these sections.

PEPPER

Rinse under running cold water, wipe dry and cut lengthwise in half, making your first incision through the centre of the stalk end (1). Remove and discard the stalk, the surrounding tough flesh, the seeds and whitish membrane. Slice lengthwise into strips as narrow or wide as required (2) and then cut these into shorter lengths or dice as required. To remove the thin, glossy outer layer of skin, spear the whole vegetable on a long-handled fork and hold over a gas burner, turning slowly, or place underneath a very hot grill and turn frequently so that the skin is evenly charred. The skin will then be easily peeled off (3) without damaging the flesh underneath; after this treatment it is usually easy to pull the stalk section out (4). Cut lengthwise in half, remove the seeds and white membrane and cut the flesh into strips if wished. Another easy way of removing the skin is to wrap each pepper tightly in foil and place in the oven, preheated to 400°F (200°C) mark 6 for about 30 minutes, turning the foil parcels over halfway through this time. Have some small sheets of newspaper ready waiting. Take the foil parcels out of the oven one at a time, unwrap the pepper and wrap it in a piece of newspaper. Leave to cool completely. When you unwrap the peppers you will find that the skin has stuck fast to the newspaper, leaving the flesh neatly exposed.

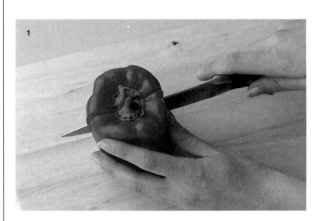

CELERY

Trim off a thin slice from the base where it is dry and discoloured, cut off the leaves and discard. Separate the stalks. Run the potato peeler down the outside of each stalk to get rid of the stringy fibres. Wash well in cold water.

ASPARAGUS

Run the potato peeler over the surface of the paler, tough ends, stripping off the outer layers, and work towards the bottom, away from the tips.
Wash the asparagus thoroughly under running cold water; trim the lower, fatter ends to roughly the same length.

PUMPKIN

Use a large, heavy kitchen knife to cut the pumpkin in half, then into quarters; cut each of these in half. Trim off any remains of stalk; scoop out the seeds and the filaments surrounding them (1). For most recipes you will have to peel the pumpkin with a medium-sized knife (2) and then dice the flesh.

ARTICHOKE

Pull off the lower, outermost leaves, which are tough (1). Place the artichoke immediately into a large bowl of cold water acidulated with the juice of 1 lemon or the cut surfaces will turn black. Cut off the stem and the top of the leaves (2) or simply snip off the pointed end of each leaf using kitchen scissors. If you are going to cook the whole, trimmed artichoke, you may choose to remove the hairy choke first; this grows out of the artichoke heart (3). Part the leaves if you have merely snipped off the tips. Very young artichokes and the smaller, tender varieties can be sliced and quartered after this preparation and all the remaining sections can be eaten but artichokes such as the Breton variety most frequently sold in Britain are best boiled whole and the leaves peeled off one by one as they only have tender flesh at the base of each leaf and at the heart. This is exposed when all the leaves have been pulled off and the choke, looking a little like an immature thistle flower, can be carefully removed with a teaspoon or a peeling knife with a curved blade to expose the 'heart' which looks like a small, thick disc slightly hollowed on top. Any remaining tough parts can be pared off. Often the stalk of very young artichokes is tender enough to cook and eat but it must first be trimmed and peeled (4).

☐ FENNEL

Slice off the tough, woody base and the top where it divides into stalks with feathery green leaves attached. Remove the outer layers. Run a potato peeler over the surface, working from the base towards the top, to remove any stringy sections on the surface (1). Part the leaves as far as you can and rinse well inside and out under running cold water. Dry and cut lengthwise into quarters. Use a small, sharp knife to trim off the woody section inside nearest the base (2). If you are going to use the fennel raw, to dip into a sauce or dressing, do not cut it up into smaller pieces. For salads, cut lengthwise into thin strips.

☐ CAULIFLOWER

Remove the outer, dark green leaves and stalks with a small, sharp knife. Cut off the stalk with a kitchen knife. If the surface of the florets has become discoloured, scrape this away with the tip of the small knife. If the cauliflower head is to be cooked whole, cut a deep cross in the base (1) and rinse in cold water before boiling or steaming. If it is to be cooked in florets, simply complete the cross cuts, slicing right through the cauliflower and dividing the head in quarters (2) and then trim off the solid section of the base, releasing the separate tender stems and their florets.

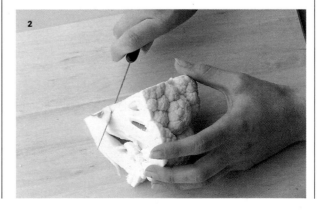

☐ CABBAGE (GREEN, WHITE OR RED), BRUSSELS SPROUTS

Remove and discard the outer, tougher or wilted leaves and slice the cabbage from top to bottom in half, then into quarters (1) with a heavy kitchen knife. Use a small, sharp knife to cut out the hard stem (2) and the largest, hardest ribs. Place the cabbage quarters in a bowl of cold water mixed with a teaspoon of vinegar, drain and cut into pieces of suitable size for your recipe. It is only necessary to trim off the base of Brussels sprouts; remove the outermost leaves if they are wilted or discoloured. Make a cross cut in the base if wished and rinse.

☐ MUSHROOMS

Brush away any compost, earth or grit with a soft brush (1) if you are using field mushrooms or other wild fungi. Cultivated mushrooms are best not washed before they are cooked; wipe them if you must with a clean, damp cloth or kitchen paper and just trim off the ends of the stalks, cutting straight across. Wild mushrooms will probably have been twisted and pulled when picked; trim the end of the stalk into a point, just enough to remove the discoloured or tough end; rinse as little as possible or their flavour and texture will suffer. Varieties whose undersides are spongy or have deep gills need very thorough rinsing in cold water.

SPINACH

If you have bought spinach that is still in little clumps, with all the stems joined at the base, remove the base by cutting the stalks about 1½–2 in (4–5 cm) above it so that you keep the tender, upper part of the stalk and the leaves (1). Rinse thoroughly in plenty of cold water. If you are serving the leaves raw as a salad, use only the tender, smaller leaves. If the spinach is to be cooked, drain briefly but do not dry and place in a saucepan with no added liquid. If you want to use the base (where most of the vitamins are concentrated), leave the spinach in clumps, trim off a thin layer from the base and cut into the remainder of the base with 3 intersecting cross-cuts (2).

CHICORY

Remove the outer leaves if they are wilted or discoloured. Use the tip of a small peeling knife to cut out a small cone-shaped piece from the base (1 & 2), leaving a narrow border around the hole as this will enable all the leaves of the chicory to stay attached. The section removed in this way is bitter and is best discarded.

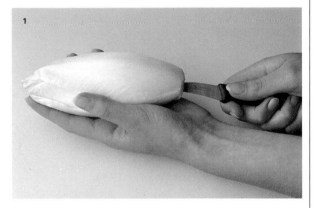

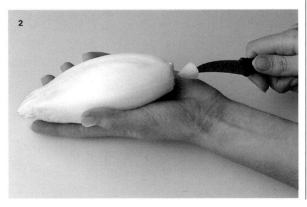

LETTUCE

Cut off the remains of the stalk to release all the leaves (1) and discard the outermost leaves and any that are wilted or discoloured. Cut the larger leaves in half, slicing through the centre rib (2), leave the smaller leaves whole. Wash all the leaves thoroughly to remove any trace of earth or small insects etc. then place in a salad spinner to dry thoroughly. If the salad is not to be served immediately, place in an airtight container in the refrigerator to keep it fresh and crisp.

RADICCHIO

Use a small peeling knife to trim the remains of the stalk to a neat point. Red Treviso chicory looks like a paler, elongated version of radicchio and is excellent cooked, but the radicchio on sale in the U.K. is best served raw in salads.

FLAT GREEN BEANS, RUNNER BEANS, MANGETOUT, FRENCH BEANS, YOUNG BROAD BEANS

Cut almost through the stalk end and pull gently towards the outer end, removing the 'string'. Trim neatly. Some of these vegetables need no more than trimming at both ends.

FRESH LEGUMES (PEAS, BROAD BEANS ETC.)

Hold the full pod in your left hand with the concave side uppermost (1); press down on the join with your thumb, making the pod split open and run your fingernail down the join to expose all the peas or beans inside. Run your thumb along the inside of the pod (2), pushing the contents out and away from you into the waiting bowl or sieve. Broad beans can have their inner skins removed as well: take each bean between your index finger and thumb and make a movement as if snapping your fingers gently; the skin will break easily, releasing the two bright green, tender cotyledons inside.

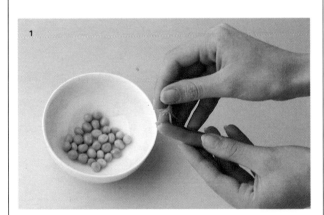

HORSERADISH

Cut off a length of root and use a very sharp peeling knife or potato peeler to remove the skin, working from the top to bottom. Grate finely; most of the flavour is contained in the thick, outer layer of the root, the core or heart is usually discarded. Wear sunglasses or, better still, goggles when grating horseradish otherwise your eyes will smart and water very badly. Freshly prepared horseradish is very strong and will dry out very quickly once grated. As it does not keep well you can grate a quantity in advance provided you place it in a jar in good vinegar; store in the refrigerator and it will be ready whenever you need it.

DRIED LEGUMES

Tiny lentils are best not pre-soaked. Just rinse thoroughly before cooking. Kidney beans of all types, chickpeas and similar dried peas, and the larger or unskinned lentils should be soaked in plenty of cold water for about 12 hours before they are cooked.

SLICING

Cutting vegetables into thin round slices is best done with a narrow-bladed knife, a filleting knife or with a mandoline slicer (see page 10, fig. 6) with an adjustable blade for different thicknesses. If using a knife, hold it so that the blade is at right angles to the chopping board, hold the vegetable firmly with your free hand, keeping your fingertips well out of the way of the knife blade (1), folding them and tucking them under as you near the end of the vegetable.

To cut the vegetable into small, rectangular pieces or julienne strips, cut the vegetable lengthwise in half, place cut side down on the board and slice again lengthwise into even thicknesses. Cut these long pieces into the required lengths (2).

Various names are given to different methods of cutting vegetables to describe size and thickness. Among the most frequently used are *chiffonade* to describe very thin strips or shreds of leaf vegetables such as lettuce, and *julienne* for very thin strips of dense vegetables such as carrots.

To dice vegetables, cut into slices of the required width, then cut across these into strips and finally into small cubes or dice (3–4).

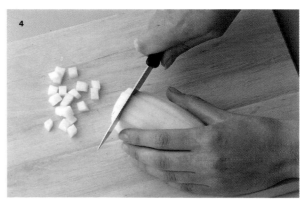

CHOPPING

A half-moon cutter (1), also known as *hachoir* or *mezzaluna* is excellent for chopping. A double-bladed version is available for chopping small quantities of herbs. Professional cooks always use large, heavy kitchen knives for chopping and practice makes them very fast and efficient: hold the handle with your fingertips rather than with the usual cutting grasp and use your free hand to hold on to the blunt edge of the blade, towards the pointed end; use this end as your pivot and move the handle up and down rapidly to chop, swinging it over the vegetables as you do so to ensure that they are all cut up finely (2).

'TURNING' OR SHAPING VEGETABLES

This term derives from the French *tourner*, trimming or cutting vegetables so that they look elegant and are of even size.

Using carrots as an example, clean and prepare medium-sized carrots as usual, then slice into sections about 2 in (5 cm) long (1); larger specimens can be cut lengthwise in half beforehand or into quarters if very large. Use the peeling knife to trim off the sharp edges at both ends, tapering the piece into an oval shape (2). Potatoes look extremely attractive when prepared this way.

25

☐ BALLING OR SCOOPING

Mainly used in vegetable cookery with large potatoes and fairly large courgettes. Hold the vegetable very firmly while you use a melon baller or scoop to cut into the vegetable; hold the scoop handle tightly while pressing down hard with one edge of its bowl and rotate the scoop as you cut into the raw potato (1) and take out a ball of its flesh; practice will make the balls neater. Scoop out as many balls as you can from each potato. Scoops are available plain or fluted and with oval bowls to cut out different decorative shapes.

☐ CANNELLING OR FLUTING

You can use a cannelle knife which has one hole or a zester which has 5 smaller holes to cut channels down the outer surface of the vegetable. Run the cannelling knife or zester down the vegetable from the top towards the stalk end while pressing firmly (1), making parallel channels in the skin and surface flesh. Slice the vegetable as usual: the slices will have attractive scalloped or deckled edges (2). Use for garnishes or salads.

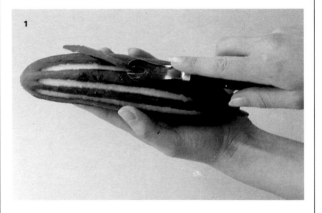

SPECIAL CUTTING METHODS FOR GARNISHES

Presentation plays an important role in serving food: decorative garnishes are pleasing to the eye and elevate any dish on to a more elegant level. All sorts of vegetables can be sculpted and cut into myriad shapes such as lattices, crabs, butterflies and many others with colour, shape and texture set off to their best advantage. Japanese and Chinese chefs in particular have turned garnishing into a minor art form.

Even when serving a meal at home, it is worth taking a little extra trouble with garnishes that are very simple but look very effective. It may just be a question of cutting a thin lemon slice in quarters and then arranging these so that their pointed sides meet to resemble butterfly wings, then place two thin strips of sweet red pepper to look like the antennae. Feather the ends of spring onions, cutting down into the stalk end towards the bulb in a series of cross cuts, soak the spring onions in iced water and they will open out like chrysanthemums or plumes.

Mangetout can be turned into small, open-mouthed fishes by the simple expedient of cutting a triangle out of one end of the flat pea pod.

When you are really short of time, a few sprigs of fresh basil or parsley will do wonders for the appearance of any dish.

Tomato rosebud
Peel off the skin of a tomato in a thin spiral with one continuous cut (1). Roll this spiral strip up to look like a rosebud (2) and complete the effect by adding a sprig of basil.

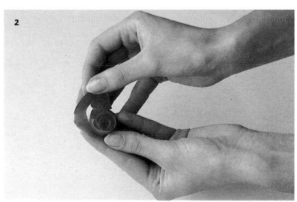

Carrot bundles

Peel the carrots, slice them lengthwise and then cut into thin strips (1).
Cut the green leaves of a spring onion into strips about ⅜ in (1 cm) wide and 5–7 in (13–18 cm) long. Blanch the carrots and green strips in boiling water for 30 seconds. Drain. Tie up small bundles of carrot strips with the green spring onion threads (2).

Radish flowers

To make a tulip, cut through the skin, tracing out 5 segments, and then work the skin away carefully, following the shape you have traced and leaving each segment attached at the bottom (1). To make a rose, slice off the very top of the radish and then cut sideways into 5 slices, leaving these attached to the base (2).

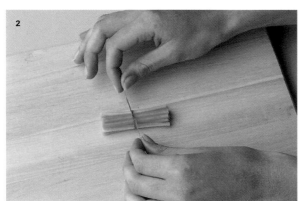

Cucumber fans

Slice a cucumber lengthwise in half. Cut off a number of 1–1¼-in (3-cm) sections, depending on how many fan garnishes you wish to make and slice diagonally into 7 flaps (1), leaving these attached at the base (originally the centre of the cucumber before it was cut in half). Fold over alternate flaps, tucking them in between the others (2).

Chilli pepper flowers

Cut in half, setting aside the pointed end for other uses (1). Use scissors to cut into 'petals' that widen towards the stalk end, stopping about ½ in (1 cm) short of the stalk base (2). Rinse under running cold water and remove the seeds. Place in a bowl of iced water and the petals will gradually open out to look like a flower.

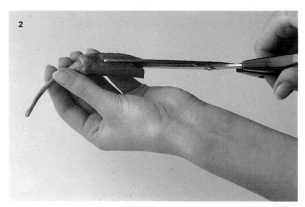

·COOKING METHODS·

Vegetables lend themselves successfully to a wide range of cooking methods. For boiling, use a stainless steel saucepan (1); a tall, narrow saucepan (2) with a removable slotted insert or basket (2) makes cooking asparagus very easy. A stainless steel or tin-lined copper fireproof casserole dish with a lid (3) is invaluable for braising vegetables. For frying you will need a non-stick frying pan (4); a wok (5) with stainless steel scoop (6) is ideal for stir-frying. A long-handled sieve (7) or wire mesh ladle makes it easy to remove vegetables from water or deep oil. A steamer (8) with slotted

steaming compartments (9) is useful. For deep-frying (10), use a thermostatically controlled electric deep-fryer with a frying basket for maximum safety. Dry-frying is easiest on a non-stick griddle (11). Shallow stainless steel pans (12) or fireproof gratin dishes (13) are suitable for browning finished dishes, while metal soufflé cases (14) or ceramic soufflé dishes or moulds (15) will be required in one or more sizes. Choose an earthenware cooking pot for lengthy simmering (16).

STEAMING

The prepared vegetables are placed in a slotted, perforated stainless steel or woven bamboo container, set over a pan containing boiling water and are cooked by the steam. Since they do not come into contact with the water at all, they lose very little of their mineral salt, vitamin content and flavour, which makes this the healthiest, most nourishing cooking method.

BOILING AND BLANCHING

The vegetables are cooked in plenty of boiling water and then drained. When vegetables are blanched they are immersed in boiling salted water for a matter of seconds or a few minutes and then drained. This makes them easier to peel, rids them of any bitter flavour where present, and can be used to seal them ready for a second, main cooking process, carried out in the oven or by frying.

STEAM-BOILING AND BRAISING

Vegetables are placed in a saucepan with a tight-fitting lid and cooked with only just enough water to keep them moist and produce steam. Some vegetables such as chicory can also be braised or steam-boiled but a little butter is needed to add extra flavour. These methods of cooking ensure that none of the taste or nutritional value of the vegetables is lost.

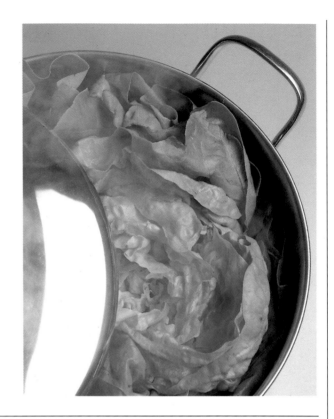

SAUTÉING, STIR-FRYING

Oil, clarified butter or fat is heated in a frying pan until it is very hot. The vegetables are then added, usually cut into small pieces so that they cook right through very quickly as they are stirred over a high heat. They remain pleasantly crisp and retain their original colour and texture. This is a method much used in Chinese cooking.

DEEP-FRYING

Oil or certain types of fat are heated to a temperature of 350°F (180°C) and the prepared vegetables are deep-fried just as they are or dipped in a frying batter. The raw vegetables may be fried in several small batches so as not to lower the temperature of the oil too much; this should remain constant and should not rise above 350°F (180°C). As the vegetables fry a thin, crisp, golden brown layer forms all over their surface.

GRILLING

Brush the vegetables with a thin film of oil then season. Place them under or over a very hot grill and cook for a very short time on one side then turn and cook on the other side. Non-stick dry-frying inserts, frying hot-plates or griddles can be used to dry-fry vegetables without any coating of oil. The heat must be sudden and intense for both methods, to make a crisp outer coating which will act as a temporary seal and prevent the moisture inside from escaping. Barbecuing is also a very successful method of grilling vegetables.

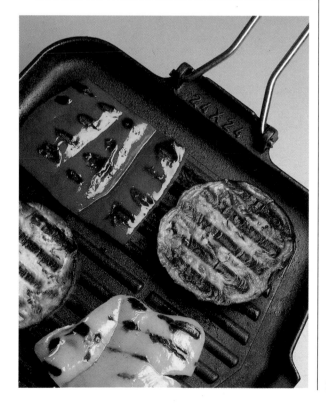

AU GRATIN

This is a finishing process for vegetables, a secondary cooking stage after they have been boiled in salted water or steamed. Place the vegetables in a single layer in an ovenproof dish greased with butter and cover them with a protective layer of béchamel or similar coating sauce, or simply dot flakes of butter over them and sprinkle with grated cheese or breadcrumbs. Place in a very hot, preheated oven or under a preheated grill and cook until the surface is crisp and golden brown.

MICROWAVE

This relatively recent addition to the choice of cooking methods means that vegetables can be cooked without any added water, fat or flavouring: the vegetables cook extremely quickly in their own moisture content and this means that little or none of their taste, texture or nutritional value is lost. To speed up cooking times still further and help the food cook evenly, a covering of cling film is stretched over the dish or plate and pierced here and there with the point of a sharp knife.

·CLASSIC SAUCES, DRESSINGS· AND STOCKS

VINAIGRETTE

Scant 2 tbsp wine vinegar

¼ tsp mustard powder

6 tbsp olive oil

salt and pepper

Optional ingredients:

¼–½ clove garlic

1 tbsp finely chopped fresh herbs or 1 generous pinch dried herbs

Mix the vinegar with the mustard powder and ½ tsp salt in a small bowl, blending thoroughly. Beat continuously as you add the oil a little at a time to ensure that it forms an evenly blended emulsion. Season with freshly ground black or white pepper to taste and use at once, just as it is or after adding extra flavouring if wished.

CITRONETTE

3 tbsp lemon juice

5 tbsp oil

salt and white pepper

Stir ½ tsp salt into the lemon juice in a small bowl until it has dissolved, then gradually beat in the oil, adding a very little at a time so that it blends in evenly. Season with freshly ground white pepper.

MAYONNAISE

3 egg yolks

1 tsp Dijon mustard

few drops Worcestershire sauce (optional)

12–14 fl oz (350–400 ml) light olive oil

1 lemon

wine vinegar

salt and pepper

Put the egg yolks in the blender with ½ tsp salt, the mustard and a few drops of Worcestershire sauce. Blend at high speed, pouring the oil very slowly through the small hole in the lid. In just over 1 minute the mayonnaise should have increased in volume and be very thick. Add the lemon juice and a few drops of wine vinegar and blend for a few seconds. Taste, and add more salt and lemon juice if needed and freshly ground white pepper to taste.

For a lighter, less rich mayonnaise, use 2 whole eggs but remember that you will have to use a blender whereas the 3-yolk quantity can be adapted to beating in the oil by hand with a whisk or wooden spoon.

VELOUTÉ SAUCE

1 oz (30 g) butter

1 oz (30 g) plain flour

16 fl oz (450 ml) chicken stock (see page 37)

salt and white pepper

The above quantities will make a fairly thick, coating sauce to cover cooked vegetables and can be used when browning them in the oven or under the grill. Bring the stock to a gentle boil. Melt the butter in a small, heavy-bottomed saucepan over a low heat. Add the flour and stir with a wooden spoon over the heat for about 2 minutes; do not allow to brown at all. Draw aside from the heat and pour in all hot stock at once, beating vigorously with a balloon whisk to ensure that the sauce is smooth and free of lumps. Return to a slightly higher heat and keep stirring with the whisk or with a wooden spoon until the sauce comes to the boil. Turn down the heat and simmer while stirring for a further minute. Take off the heat and stir in salt and freshly ground pepper to taste.

BÉCHAMEL SAUCE

1 carrot

½ celery stalk

1 onion stuck with 2 cloves

few peppercorns

1 bay leaf

16 fl oz (450 ml) milk

1 oz (30 g) butter

1 oz (30 g) plain flour

nutmeg

salt and white pepper

This is a useful sauce for coating vegetables, particularly prior to browning in the oven or under the grill in gratin dishes. Bring the milk to the boil with the carrot, celery, onion, bay leaf and peppercorns. Remove from the heat and leave to infuse for 30 minutes. Remove and discard the vegetables. Melt the butter in a small pan, add the flour and stir for 1–2 minutes. Gradually add the flavoured milk. Bring to the boil, stirring continuously, and cook for 1–2 minutes until thick. Season with nutmeg, salt and white pepper.

CRÊPES OR PANCAKES

10 fl oz (270 ml) cold milk

8 oz (220 g) sifted plain flour

4 eggs

4 tbsp melted butter

2 tbsp oil

salt

The above quantities will yield enough batter for 18 crêpes or pancakes, each about 7 in (18 cm) in diameter. Pour the milk in the blender; add the same volume of iced water and add the sifted flour with ½ tsp salt. Break in the eggs and add the butter. Cover and blend at high speed for 1 minute. Uncover and scrape off the flour that has stuck to the sides, replace the lid and blend for another ½ minute. Chill the batter in the refrigerator for at least 6 hours before using.

Lightly oil the inside of a non-stick frying pan before cooking the first pancake (there is no need to oil again). Heat the pan over a moderately high heat; when very hot pour 3 tbsp of the batter into the centre of the frying pan and immediately tip the pan this way and that so that the batter spreads out to coat the entire bottom of the frying pan thinly. Use a non-stick spatula to loosen the edges of the pancake, turn or toss the pancake when the first side is pale golden brown and cook the other side, then slide out of the pan on to a heated plate.

Repeat this process with the remaining batter, stacking the cooked pancakes one on top of another. The pancakes can be made in advance and kept in the refrigerator, covered with cling film; they also freeze extremely well.

CLARIFIED BUTTER (GHEE)

Melt the butter over a low heat in a small, heavy-bottomed saucepan and leave over this gentle heat for 10 minutes to allow the white whey or casein to coagulate at the bottom of the pan. Draw aside from the heat. Leave to stand for a few minutes and then pour through a sieve lined with a piece of muslin into a storage jar. Seal very tightly and store in the refrigerator or, for shorter periods, at room temperature. Indian recipes frequently call for *ghee* and it is also used in other cuisines for frying at temperatures at which unclarified butter would burn since it is the casein content that burns.

COURT-BOUILLON

3 shallots, peeled and sliced

20 parsley stalks, cut in sections

2 sprigs fresh thyme

1 bay leaf

18 fl oz (500 ml) white wine vinegar

3½ fl oz (100 ml) best virgin olive oil

salt and 5 each black and white peppercorns

This *court-bouillon* is suitable for pickling vegetables or for cooking ready for bottling in oil. For a straightforward aromatic cooking broth, use 2 fl oz (50 ml) fresh, strained lemon juice or 2 tbsp white wine vinegar for every ¾ pint (450 ml) cold water for the liquid content, otherwise the flavour of the *court-bouillon* will be too sharp.

Place all the listed ingredients in a large saucepan with 14 fl oz (400 ml) cold water (tie the thyme and peppercorns in a piece of muslin). Add 1 tbsp coarse sea salt. Bring slowly to the boil over a very gentle heat, add the vegetables and simmer until they are tender but still have a little bite left in them. If they are to be bottled, drain and arrange in layers in suitable jars and completely cover with olive oil. Seal tightly and keep in a store cupboard at room temperature or in the larder.

BEEF STOCK

1½ lb (700 g) lean stewing or braising beef

1 medium-sized leek

1 small onion

1 medium-sized carrot

1 baby turnip

1 green celery stalk with its leaves

few parsley stalks

1 bay leaf

2 cloves

salt and black peppercorns

Prepare the vegetables and rinse them well. Make a deep cross cut in both ends of the onion and stud it with the cloves. Place all the ingredients in a very large saucepan or stockpot with 3½ pints (2 litres) cold water, using 5 black peppercorns and 1 tbsp coarse sea salt. Cover and bring up to boiling point, skimming the surface as the scum rises. Simmer very gently for 3 hours.

Remove the beef; strain the stock through a piece of muslin placed in a sieve or through a very fine sieve held over a large bowl. Add a little more salt if necessary and when cold, place in the refrigerator to chill overnight. Take the stock out of the refrigerator and immediately remove the layer of solidified fat from the surface. The beef will not be wasted: eat it hot with a sharp, vinegary sauce or serve it cold the next day, thinly sliced with chutney or pickles.

CHICKEN STOCK

1 chicken carcass

1 leek

2 carrots

1 onion

1 celery stalk

few parsley stalks

1 sprig thyme

few black peppercorns

salt

Break up the carcass and place, together with any leftover trimmings, in a large pot. Add all the other ingredients and pour in 3 pints cold water. Cover and bring to the boil. Skim off any scum then simmer gently for 2 hours. Strain the stock through muslin or a fine sieve and cool. Chill in the refrigerator and remove the layer of solid fat from the surface before using.

—•—

VEGETABLE STOCK

2 celery stalks with leaves

2 carrots

1 large onion

1 leek

1 potato

few parsley stalks

1 bay leaf

few peppercorns

salt

Finely chop the vegetables and place in a large saucepan with the other ingredients. Pour in 1½ pints cold water, cover and bring to the boil. Simmer for 15 minutes then strain through a fine sieve.

—•—

FISH STOCK (FUMET)

2–2½ lb (1 kg) fresh fish trimmings, heads, bones etc.

3 large shallots or 1 medium-sized mild onion

1 small carrot

2 heads celery

2 sprigs fresh thyme

1 bay leaf

18 fl oz (500 ml) dry white wine

salt and 5 each white and black peppercorns

Rinse the fish trimmings well under running cold water, break or cut up any very large pieces and place in a very big saucepan or stockpot with 2¾ pints (1½ litres) cold water and ½ tsp salt. Bring

slowly to the boil over a gentle heat, skimming off any scum as it rises to the surface. Peel the shallot and cut into thin rings; slice the peeled carrot and the celery finely and cut the celery stalks into smaller pieces. Add these to the water and fish and simmer uncovered for 30 minutes.

Rinse a clean, close-weave cloth in cold water and wring out; use it to line a sieve or colander and slowly pour the fish stock through it into a large bowl below to strain. Use immediately or refrigerate for up to 2 days.

—•—

SHELLFISH STOCK

2–2¼ lb (1 kg) prawn heads and shells

1 large spring onion

1 medium-sized carrot

1 small celery stalk

20 parsley stalks (without leaves)

1 small green chilli pepper (optional)

2 sprigs fresh thyme

1 bay leaf

small pinch fennel seeds (optional)

¼ tsp mild curry powder (optional)

7 fl oz (200 ml) dry vermouth or dry white wine

salt and 10 white peppercorns

Trim the spring onion, carrot and celery, peeling where necessary, and slice finely; cut the parsley stalks into small pieces; remove the seeds and stalk from the green chilli pepper and slice into thin rings. Place all the ingredients in a very large saucepan or stockpot together with 1¾ pints (1 litre) cold water and ¼ tbsp coarse sea salt. Bring to the boil very slowly over a gentle heat and simmer gently, uncovered, for 20 minutes, stirring from time to time. Strain through a large, fine mesh sieve, crushing the prawn heads down hard with the back of a large wooden spoon to squeeze out all the juices.

·Storing and Preserving·

There is nothing to equal freshly picked vegetables, but you can ensure plenty of taste and vitamin content if you buy the vegetables you require as frequently as possible and store for the minimum length of time in the vegetable compartment of your refrigerator. Supermarkets have found the best way of keeping vegetables in moderately good condition for several days by arranging them in single layers in polystyrene trays, covered with clingfilm. Even when cooked, vegetables should be refrigerated in airtight containers or they will deteriorate.

Deep-frozen vegetables will, of course, keep for months and many will lose little of their taste and texture.

Anyone who has a vegetable garden will sometimes find that there is a glut of certain vegetables and these are best deep-frozen to avoid wastage or the prospect of serving them with monotonous regularity. Some vegetables can be frozen raw (peeled, chopped onion, celery, mushrooms, tomatoes, peppers and herbs) but most need brief pre-cooking or blanching for about 5 minutes. They are then transferred to a sieve or colander and plunged into iced water to cool quickly. After a thorough draining and drying they can be packed in special foil containers or polythene bags with all or most of the air removed from inside by sucking it out with a straw inserted into the bag, held gathered tightly around it and quickly sealed, labelled with the contents and freezing date, then quick-frozen in the freezer. They should be eaten within 8 months, preferably sooner.

Deep-freezing raw vegetables

Mushrooms: for cultivated mushrooms, simply trim off the ends of the stalks and wipe with a damp cloth. Wild mushrooms: brush to get rid of grit, scrape off any discoloured sections with a peeling knife and trim off the root end. Place in polythene bags, seal tightly after extracting as much air as possible, label and date. Use within 3 months.

Tomatoes: choose firm, ripe tomatoes. Wash and dry them thoroughly. Place in a single, well spaced layer on a polystyrene tray and open-freeze; transfer them to a polythene bag, seal, label and date. Use within 8 months of freezing date.

Peppers: wash, dry, cut in half and remove the stem, seeds and white membrane inside. Open-freeze and store as for tomatoes.

Celery: cut off the base and run a swivel-blade potato peeler along the outer surfaces to get rid of any strings; brush or wipe off any traces of soil. Freeze the stalks whole and use only for cooked dishes (casseroles, flavourings for sauces etc.).

It is better not to thaw vegetables before cooking them but to cook them from frozen; if you must thaw them, leave them to defrost inside their unopened packaging until just before using them as they deteriorate rapidly once exposed to the air.

Deep-frozen cooked vegetable dishes can be placed straight in pre-heated conventional, convection ovens and, of course in microwave ovens and reheated from frozen provided you have used containers that are safe in the freezer and in the oven or microwave.

• Wash vegetables thoroughly but quickly; do not leave them soaking in water for a long time. Cook them in as little water as possible to prevent excessive loss of mineral salts and vitamins.

• When you have used water for boiling vegetables, keep it for making delicious soups or sauces, or add to casseroles, risottos and similar dishes; vegetable stock is full of precious nutrition and too good to waste.

• Once trimmed and sliced, the cut surfaces of globe artichokes discolour and darken as oxidization takes place; to prevent this happening, drop them into a bowl containing some cold water acidulated with lemon juice immediately after slicing or cutting them.

• The smell of onions and garlic is penetrating and persistent; keep a separate, small chopping board exclusively for this purpose. The only totally effective method to prevent your eyes watering while slicing onions or grating horseradish is to wear goggles!

• Never slice or chop your onions until just before you intend cooking them; if chopped and left for a long time they acquire an unpleasant smell. You can, however, chop them (on their own or mixed with chopped carrot and celery as a ready-to-use mirepoix mixture) and store straight away in the freezer, either in an ice container designed for the purpose or in the ice cube tray; take out the cubes when frozen solid and store in polythene bags. Take out the number of cubes you need as and when you need them.

• It is always a good idea to keep plenty of chopped parsley in the freezer: store it in a sealed container and take out as much as you need at a time, breaking and crumbling it off the block and putting the rest back in the freezer.

• Use the stalks as well as the leaves of spinach; much of the mineral salts and vitamins are concentrated towards the lower ends of the stalks; save the stalks and use them for making soups or omelettes.

• To get rid of the bitterness sometimes present in aubergines and cucumbers (although most of today's varieties have been cultivated to minimize this), slice and spread out on a clean cloth or kitchen paper in a single layer, sprinkle with salt and leave for 30 minutes to 1 hour, then cover with another cloth and press down firmly to absorb the liquid. Alternatively, place the slices in a colander, sprinkling each layer with a little salt and place a plate and a weight on top; dry the slices with kitchen paper.

• When dicing or cubing aubergines for use in vegetable casseroles, leave the skin on so that the vegetable is recognizable and gives the dish extra colour and interest.

• When removing seeds from tomato slices or when hollowing out whole tomatoes ready for stuffing, keep all the seeds, caps and pieces of flesh; sieve them and use the liquid to flavour sauces and soups.

• Cauliflower and cabbage produce an unpleasant, pervasive smell when cooking due to their sulphur content; to lessen this add a few drops of vinegar or a cube of stale bread to the cooking water. Both methods also help to prevent red cabbage turning a bluey-purple colour when cooked.

• To preserve the bright green colour of certain vegetables, such as French beans, and prevent them turning a darker, dull colour when done, cook until just tender but still crisp in an uncovered pan containing plenty of lightly salted water. Many chefs refresh the crisp vegetables under cold water and then steam to reheat which is effective, and also practical when serving large numbers but results in more mineral salts and vitamins being lost.

• Nettles must always be blanched before use to eliminate any possible toxicity.

·LEAF AND STALK VEGETABLES·

*Pasta with
turnip tops*

Cabbage and ham rolls

Fennel and lettuce au gratin

CELERY AND GREEN APPLE JUICE

4 crisp green celery stalks

4 green dessert apples

1½ pints (900 ml) unsweetened apple juice

Chop the celery. Wash, quarter and core the apples but do not peel them; cut the quarters into fairly small pieces. Pour the apple juice into the blender, add the celery and apples and process at high speed until smooth. Unless you have a large capacity blender or liquidizer, you will need to process the ingredients in 2 batches. If the juice is too tart for your taste, add a generous pinch of sugar. Alternatively, add a small pinch of celery salt and a dash of Tabasco. Serve with ice cubes.

———— • ————

HAM AND CRESS SANDWICHES

4–5 boxes mustard and cress or 1–2 bunches watercress

8 thin slices white or brown bread

7–8 oz (200–225 g) very thinly sliced ham or cold roast pork

softened butter

Dijon mustard

salt and pepper

Rinse the cress thoroughly in a sieve, spread out between kitchen paper to dry or use a salad spinner. If using watercress, take all the leaves off the stalks. Trim off the crusts from the bread. Spread 4 slices generously with butter, the other 4 with a very thin layer of mustard. Place the ham or pork on the 4 slices spread with mustard and season lightly with salt and freshly ground pepper. Place plenty of cress on top of the meat, heaping it up in the centre. Sprinkle with a pinch of salt and cover with the buttered slices, pressing down firmly.

INDIAN BREAD WITH SAVOY CABBAGE STUFFING

Paratha

9 oz (250 g) wholemeal flour

4 fl oz (120 ml) ghee (clarified butter, see page 36)

8 oz (230 g) shredded Savoy cabbage

4 oz (100–120 g) fresh curd cheese

½–1 small green chilli pepper, finely chopped

3 tbsp chopped coriander leaves

salt

Sift the flour into a large bowl with ½ tsp salt, stir in scant 2 tbsp of the clarified butter and gradually add 6–8 fl oz (200–225 ml) cold water; the dough should be soft but not too moist to knead easily. Knead until smooth and fairly elastic, shape into a ball, wrap in foil or cling film and chill for 20 minutes.

Cut out all the large ribs from the cabbage; wash the leaves, dry thoroughly and shred. Place in a large bowl and sprinkle with 2 tsp salt. Stir, then leave to stand for 15 minutes. Transfer to a sieve or colander and rinse thoroughly under running cold water, drain and squeeze out excess moisture. Place in a bowl and mix with the crumbled fresh cheese, the chillis, coriander and a little salt. Divide the dough into 10 evenly sized pieces; roll each one into a ball between your palms keeping the ones you are not working on covered with foil or cling film as they dry out quickly. Roll out these balls one at a time with a rolling pin into thin discs about 4½–5 in (12 cm) in diameter. Place equal amounts of the cabbage and cheese stuffing on 5 discs, cover with the remaining 5 and seal the edges by pinching together (moisten the edges if necessary). These *parathas* must be fried at once: lightly coat the inside of a large cast iron frying pan with ghee and heat until very hot; cook the parathas for 2 minutes on each side. Keep the fried parathas hot in the oven, uncovered, while you finish cooking the rest.

———— • ————

SALAD HORS D'OEUVRE WITH POMEGRANATE SEEDS AND FOIE GRAS

1 ripe pomegranate

8 oz (220 g) small-leaved salad greens

4 medium-sized lettuce leaves

4 slices bread

8 butter curls

1 lemon

12 thin slices rare beef fillet

4 slices tinned truffle-flavoured foie gras

light olive oil

salt

Peel the pomegranate and take out the seeds. Cut the bread slices diagonally in half, into two triangles, and toast them lightly. Use a butter curler or an ordinary knife to cut the butter curls out of very firm butter and place them in a small bowl of iced water. Just before serving, mix a simple dressing with 3 tbsp olive oil, 1–1½ tsp lemon juice and a small pinch of salt, pour over the salad greens and toss until evenly mixed. Heap an equal quantity of salad in the centre of 4 plates and place 3 slices of beef on top, sprinkle the pomegranate seeds on top.

Place a lettuce leaf beside each of these mounds, and put a slice of *foie gras* and 2 butter curls on each leaf. Lastly, arrange 2 slices of toast to one side of the *foie gras*. Drizzle 1–1½ tsp olive oil over each portion of beef and pomegranate seeds and serve.

·•·

ITALIAN EASTER PIE
Torta pasqualina

For the pastry:

10½ oz (300 g) strong plain flour

3½ fl oz (100 ml) olive oil

For the filling:

1 lb (500 g) spinach

1 lb (500 g) Swiss chard

1 medium-sized onion

1 oz (30 g) butter

9 very fresh small eggs

6 tbsp grated Parmesan cheese

1 small clove garlic, finely chopped

1 tbsp finely chopped fresh marjoram

9 oz (250 g) ricotta cheese

extra olive oil for greasing

salt and pepper

Sift the flour with a generous pinch of salt into a large mixing bowl. Make a well in the centre and pour 7 fl oz (200 ml) boiling water and the oil into it. Gradually stir all this liquid into the flour then knead for about 10 minutes, until the dough is smooth and homogeneous. Divide into 6 pieces of equal size, shape each into a ball; place these in a single layer in the mixing bowl and cover with a damp cloth. Leave for 1 hour.

While the pastry dough is resting, make the filling: remove the stalks and wash the spinach and Swiss chard well and blanch both in a large pan of boiling, salted water for 5 minutes. Drain, leave to cool, then squeeze out all the moisture and chop finely. Peel the onion and chop very finely; sweat in the butter with 1 tbsp olive oil in a covered saucepan over a low heat until wilted and tender. Add the chopped spinach and chard leaves and cook, uncovered, for 5 minutes, stirring continuously with a wooden spoon to allow the moisture to evaporate. Remove from the heat and leave to cool.

Break 5 eggs into a large mixing bowl and beat lightly, seasoning with a little salt and freshly ground pepper. Stir in the Parmesan cheese, the garlic, marjoram, ricotta and the spinach mixture. Taste and correct the seasoning as required. Preheat the oven to 475°F (240°C) mark 9.

Grease a 10–10½-in (26-cm) fairly shallow cake tin with a little olive oil. Take a dough ball out of the bowl, keeping the rest covered. Flatten it with the palms of your hands, carefully pushing and pulling it into a very large, thin circle and use this to line the base of the tin, allowing for an extra ¾ in (2 cm) up the sides. Brush lightly with a pastry brush dipped in olive oil.

Repeat this operation with a second dough ball, placing the circle on top of the first one, pressing the spare pastry against the sides of the cake tin and brushing its surface with oil. Do the same with a third ball, but do not brush the surface with oil. Spread all the filling in a very thick, even layer on top of the third pastry layer. Press the back of a large wooden spoon gently down into the filling to make 4 evenly spaced shallow hollows and break an egg into each one. Season each egg with a little salt and freshly ground pepper. Make another circle out of the fourth dough ball, place it carefully on top of the eggs and filling and press the margin against the sides of the cake tin just as you did before but insert a drinking straw at one point as you do so. Brush the surface very lightly with oil. Repeat this process with the last two portions of pastry dough, but do not oil the surface of the last, topmost layer. Blow gently but steadily down the straw to introduce as much air as possible between the filling and the layered pastry lid. Seal the little aperture by pressing it against the inside of the cake tin as you remove the straw. Run a knife between the inside of the tin and the upper edges of the pastry layers, then roll these over, away from the edge of the tin; pinch them to seal firmly. Bake for 30 minutes.

———— • ————

STUFFED CHINESE LEAF ROLLS

8 large Chinese leaves

2 large leaves nori *(black Japanese seaweed sheets)*

4 oz (120 g) boneless chicken breast, skinned

4 oz (120 g) cooked ham

pinch monosodium glutamate (optional)

1 tbsp Chinese rice wine or dry sherry

salt and white pepper

For the sauce:

9 fl oz (250 ml) chicken stock (see page 37)

1 tbsp cornflour or potato flour

pinch ground ginger

1 tbsp Chinese rice wine or dry sherry

salt and white pepper

Chop the chicken and ham finely and transfer to a bowl. Season with a pinch of salt, a small pinch of monosodium glutamate if wished, a pinch of pepper and 1 tbsp rice wine or sherry; stir well.

Blanch the Chinese leaves one at a time for 30 seconds in a large pan of boiling salted water, spreading out each one flat to dry on kitchen paper as you take it out of the water. Cut a V-shaped section out of bottom of the stalk of each leaf with a small, sharp knife, removing enough to make rolling the leaves easy when it comes to wrapping the stuffing. Pile 4 leaves one on top of the other 'head to tail' (this will prevent the filling oozing out); do the same with the remaining 4 leaves in a separate pile. Sprinkle a thin layer of cornflour or potato flour over the exposed surface of the 2 topmost leaves, spread out a piece of nori seaweed on top of both (this can be bought at most Oriental grocers) and then cover each with half the filling spread out in a thick layer. Carefully roll up the leaves and filling into 2 long sausage shapes; they will not be very neat, but this does not matter. Transfer these to a lightly oiled heatproof plate and steam over fast boiling water for 20 minutes.

Meanwhile, make the sauce. Mix the cornflour or potato flour with 1½ tbsp cold water and stir into a saucepan containing the chicken stock. Season with a pinch each of salt and pepper, the ginger, rice wine or dry sherry. Bring to the boil, stirring continuously with a balloon whisk or wooden spoon. Simmer for 1–2 minutes until the sauce has thickened. When the rolls have steamed for 20 minutes, cut them into slices ¾ in (2 cm) thick. Hand round the sauce separately.

———— • ————

CELERY CANAPÉS WITH ROQUEFORT STUFFING

1 head white celery
7 oz (200 g) Roquefort cheese
5 oz (150 g) soft cream cheese
salt and pepper

Separate the celery stalks and cut off their tapering ends. Rinse thoroughly, dry then run a potato peeler over their outer sides to remove any strings. Cut the stalks into sections about 1½ in (4 cm) long. Make the filling: use a fork to break up the Roquefort and work it into a paste; gradually work in the cream cheese until smoothly blended. Season to taste with salt and freshly ground white pepper.
Fill the celery canapés with this mixture. Cover with cling film and refrigerate until required.

— • —

CAESAR SALAD WITH QUAIL'S EGGS

1 Cos or iceberg lettuce
5 tinned anchovy fillets
6 small slices brown bread
olive oil
24 cooked quail's eggs, shelled
for the anchovy-flavoured citronette dressing:
½ clove garlic
2 tbsp lemon juice
½ tinned anchovy fillet, very finely chopped
2 oz (50 g) freshly grated Parmesan cheese
6 tbsp extra-virgin olive oil
freshly ground pepper

Wash the lettuce, drain and dry, then shred into very thin strips or simply tear the leaves into small pieces. Chop the anchovy fillets. Make croutons by cutting the bread into small squares and frying these in a little olive oil in a non-stick frying pan; sprinkle them with a little salt and set aside. Place all the dressing ingredients in a blender and process until smooth. Just before serving, spread out the shredded lettuce on 4 individual plates, space out an equal number of anchovy pieces, quail's eggs and croutons on top of each bed of lettuce and sprinkle with the citronette dressing. It is unlikely that you will want to add salt as the anchovies and Parmesan both add a salty tang to the salad.

— • —

SALADE GOURMANDE WITH SHERRY VINAIGRETTE

2 small heads chicory
1 small head radicchio
7 oz (200 g) Webb's Wonder or Cos lettuce (heart only)
7 oz (200 g) cold cooked French beans
4 slices tinned truffled foie gras, chilled
For the sherry vinaigrette:
2 tbsp sherry vinegar
6 tbsp olive oil
1 tsp Dijon mustard
salt and pepper
Serve with:
hot French bread or toast

Trim, rinse and drain all the salad leaves. Dry with kitchen paper or in the salad spinner. Slice the chicory lengthwise into thin strips; cut or tear the larger leaves of the radicchio and the round lettuce heart into small pieces. Run a butter curler over the chilled *foie gras* very lightly to take off thin, curling strips, rather like butter curls. Mix the salad leaves with the French beans in a wide salad bowl.

Make the sherry vinaigrette: blend the vinegar, mustard and salt in a small bowl using a wooden spoon or birch whisk; gradually add the olive oil, beating continuously. Stir in freshly ground white pepper. Dress the salad with this vinaigrette, making sure every leaf is coated; transfer an equal quantity of the dressed salad to 4 plates; place an equal number of *foie gras* curls on top and serve at once with hot French bread.

———•———

Aida Salad

2 heads curly endive or frisée
2 hard-boiled eggs
2 red peppers
1 melon
2 cooked or tinned artichoke hearts
citronette dressing (see page 35)

Use only the paler, inner sections or hearts of the curly endives; wash the leaves thoroughly, drain and dry well, then tear into fairly small pieces. Push the hard-boiled eggs through a sieve into a bowl, forcing them through with the back of a large wooden spoon. Place the red peppers under a hot grill, turning frequently, to loosen their skins. Remove the skins and cut the peppers in half, removing the stalk, seeds and white membrane. Slice into short, thin strips. Cut the melon in half, remove the seeds and use a melon baller to scoop out small balls of flesh. If you are using fresh artichokes, drop into a bowl of cold water to which the juice of ½ lemon has been added as you prepare and slice the hearts (see page 18) or they will discolour; Drain the artichoke hearts at the last minute and cut into strips.

Arrange a bed of curly endive on each plate, sprinkle with the artichoke strips, place the melon balls in the centre, surrounded by the red pepper strips and sprinkle with the sieved hard-boiled egg. Hand round the citronette dressing separately.

Chicory and Prawn Salad

3–4 heads chicory
14 oz (400 g) tiger prawns
For the walnut citronette dressing:
2 oz (60 g) shelled walnuts
3 tbsp lemon juice
3½ fl oz (100 ml) olive oil
1 tsp walnut oil (optional)
pinch caster sugar
salt and white pepper

If you have bought raw prawns, steam them for 3–4 minutes, then peel them. Prepare the chicory, rinse briefly and dry. Carefully remove 12 undamaged outer leaves and set aside.

Make the sauce: Add ½ oz (10 g) of the walnuts to boiling water and simmer for 2 minutes, drain and peel off the thin inner skin if present. Chop coarsely and place in the blender with the lemon juice, freshly ground pepper and 1 tsp salt. Process at high speed for a few seconds, then continue to blend while trickling in the oil. Pour into a small jug or sauceboat. Chop the remaining walnuts very coarsely and set aside.

Fan out 3 of the reserved outer leaves on each of the 4 plates, to one side. Shred the remaining chicory leaves into a chiffonade (see page 24) and place in the centre of each plate; arrange an equal number of prawns on each bed and sprinkle with some of the dressing. Decorate with the chopped walnuts and serve at once.

———•———

Turkish Stuffed Vine Leaves

40 small vine leaves, fresh or vacuum-packed
7 oz (200 g) onions
1 oz (30 g) pine nuts

3½ oz (100 g) easy-cook rice
generous pinch ground cinnamon
1 oz (30 g) sultanas, soaked in lukewarm water
2 lemons
olive oil
salt
2 dried allspice berries, crushed

These quantities should yield about 30 stuffed rolls, allowing for a few spoiled ones. Preheat the oven to 300°F (150°C) mark 2. Chop the onion finely. Heat 4 tbsp olive oil in a medium-sized saucepan and fry the onion gently, stirring frequently until lightly browned, then add the pine nuts and cook over a low heat for 2 minutes, stirring.

Add the rice, scant ½ tsp salt, the cinnamon and the allspice. Drain the sultanas and squeeze out excess moisture, add to the saucepan and continue cooking and stirring for 1 minute more.

Add 9 fl oz (250 ml) boiling water to the rice and other ingredients. Cover the saucepan tightly, turn down the heat to very low and cook for 15 minutes or until all the water has been absorbed and the rice is done. Stir in a little more salt if needed and leave to cool. If using fresh vine leaves, wash them and blanch for 30 seconds in a large pan of boiling water. Drain and spread them out carefully to lie flat, veined side downwards, on kitchen paper. If you are using vacuum-packed vine leaves, just rinse and drain well and spread out.

Grease a wide, fairly shallow ovenproof dish with oil and spread out 10 of the vine leaves to cover the bottom. Place 1–1½ tsp of the rice mixture towards the stem end of each leaf and roll up into neat parcels tucking in the spare uncovered leaf space on either side as you do so, to prevent the rice filling from seeping out. Pack into the dish, preferably in a single layer. Sprinkle with 2 tbsp oil and 4 tbsp cold water. Cover the dish with a sheet of foil and bake in the oven for 50 minutes. Serve warm or cold as a starter, with lemon wedges for each person to squeeze over them.

CHICORY, FENNEL AND PEAR SALAD

4 oz (120 g) rocket
4 oz (120 g) lamb's lettuce or substitute (see method)
2 heads chicory
2 small bulbs fennel
2 tomatoes
16 shelled walnuts, chopped
2 pears (Comice if available)
1 lemon
mustard-flavoured citronette dressing (see page 78, half quantity)

Trim, wash and dry all the vegetables; mix the salad greens together in a large bowl. If you can find neither rocket nor lamb's lettuce, almost any other salad greens may be substituted: try curly endive, oak leaf lettuce or lollo rosso. Slice the chicory lengthwise in half, cut across the middle of these halves, then cut into julienne strips (see page 24); repeat this operation with the fennel bulbs, having first removed their outermost layer and mix them together in another bowl. Slice each of the tomatoes into quarters or 8 wedges. Peel the pears, cut them lengthwise into quarters, remove the cores and slice; sprinkle with lemon juice.

Arrange all these prepared ingredients attractively in a bowl or in a large fan shape on a serving platter and sprinkle with the walnuts. Serve with the mustard-flavoured dressing.

——— • ———

GREEN SALAD WITH ROQUEFORT DRESSING

2 lettuce hearts
For the Roquefort dressing:
2 tbsp wine vinegar

1½–2 oz (40–50 g) Roquefort cheese, crumbled

6 tbsp light olive oil

2 tbsp single cream

salt and freshly ground pepper

Separate the lettuce heart leaves, removing the hard stalk end. Wash and spin, shake or pat dry. Tear the larger, outer leaves of each heart across in half and place on 4 individual plates or in a large bowl. Place the small leaves on top. Make the dressing. Mix ½ tsp salt with the vinegar in a bowl. Add the crumbled Roquefort and work into the vinegar with a fork until fairly smooth. Mix in 1 tbsp cold water and some freshly ground pepper. Add the oil in a thin trickle while beating with a balloon whisk or birch whisk. Beat in the cream. Sprinkle the Roquefort dressing over the lettuce and serve at once.

———•———

FENNEL, ORANGE AND WALNUT SALAD

4 bulbs fennel

2 large oranges

8 shelled walnuts

extra-virgin olive oil

salt and white pepper

Remove and discard the outermost layer from the fennel bulbs. Trim, wash and dry the bulbs and cut them lengthwise into quarters, then cut these pieces into matchstick strips.
Arrange the fennel strips on individual plates in a fan shape; sprinkle with a little salt, freshly ground white pepper and a little olive oil.
Peel the oranges with a knife, cutting away the inner skin as well as the peel and pith; take the segments out of their thin membranous pockets, remove any seeds and fan out the segments on top of the fennel. Reserve 4 walnuts for decoration; chop the others

coarsely and sprinkle over the orange segments. Place the reserved walnuts at the pointed, lower end of each fan and serve.

———•———

CHICORY SALAD WITH PORT, SULTANAS AND WALNUTS

4 medium-sized heads chicory

3 tbsp sultanas

3½ fl oz (100 ml) port

6 walnuts

citronette dressing (see page 35)

Soak the sultanas in the port until soft and plump. Rinse the chicory briefly and dry thoroughly. Cut off the tough end then slice lengthwise in quarters; cut these pieces into 1½-in (4-cm) lengths. Chop the walnuts coarsely.
Arrange the salad in individual bowls, sprinkle with the drained port-flavoured sultanas. Sprinkle the citronette dressing all over the salad and serve at once.

———•———

SPINACH SALAD WITH SULTANAS AND PINE NUTS

1–1¼ lb (450–600 g) spinach

2 tbsp sultanas

2 tbsp pine nuts

For the dressing:

2 tbsp lemon juice

6 tbsp olive or sunflower oil

salt and white pepper

Pick out all the youngest, crispest spinach leaves with thin, tender stalks; reserve the larger leaves for use in other dishes. Wash these small, raw leaves very well in several changes of water and dry in a salad spinner. If the sultanas are rather firm and dry, soak briefly in warm water to soften and plump up; drain and squeeze out excess moisture. Arrange the spinach leaves on 4 individual plates and sprinkle sultanas and pine nuts over each portion.

Mix ½ tsp salt with the lemon juice until it has dissolved. Beat in the oil a little at a time. Season with freshly ground pepper and sprinkle over the salad.

———•———

PERSIAN SPINACH WITH YOGHURT DRESSING
Borani

1¼ lb (600 g) very fresh young spinach (net prepared weight)

1 medium-sized shallot or ½ small onion

9 fl oz (250 ml) Greek yoghurt

2 tbsp fresh lime juice

2 tsp chopped fresh spearmint or apple mint

salt and black pepper

Bring a large saucepan of salted water to the boil. Trim and wash the spinach thoroughly, selecting the smaller leaves with thin stalks; the large leaves and stalks can be used for other dishes, such as soup. When you are sure you have removed every trace of grit, add the leaves to the saucepan of boiling water and blanch for 2 minutes. Drain well, leave to cool and then squeeze out as much moisture as possible. Chop very finely and place in a serving bowl.

Chop the peeled shallot or mild onion very finely and mix with the yoghurt in a small bowl, adding ½ tsp salt, the lime juice and a little freshly ground pepper. Stir this yoghurt dressing into the spinach, cover and chill for 2 hours in the refrigerator. Just before serving sprinkle with the mint.

SPINACH JAPANESE STYLE
Horenso no Ohitashi

2–2½ lb (1 kg) very fresh spinach

6 tbsp Japanese soy sauce

½ tsp caster sugar

pinch ajinomoto (Japanese taste powder) or monosodium glutamate (optional)

4 tbsp katzuobushi (Japanese fish scales) or 4 tbsp toasted sesame seeds

The Japanese ingredients in this recipe are available from Oriental grocery shops and larger delicatessens and health food shops. Trim the spinach leaves, picking out the small leaves with thin, tender stalks and wash them thoroughly. You should allow about 1¼ lb for this dish. Blanch in a large saucepan of boiling salted water for a maximum of 2 minutes. Drain very thoroughly and leave to cool. Do not squeeze out the moisture.

Make the sauce: mix the soy sauce in a small saucepan with 2 tbsp water, the sugar and the taste powder if used. Heat gently, stirring, until the sugar has completely dissolved. Leave to cool.

Arrange the spinach on 4 individual plates, moisten with the soy dressing and sprinkle with the fish scales or sesame seeds.

———•———

SPINACH MOULDS

9–10 oz (250–350 g) spinach leaves (net prepared, precooked weight, see method)

1 shallot or ½ mild red onion

scant ½ oz (10 g) butter

4 large eggs

5 fl oz (150 ml) double cream

5 fl oz (150 ml) milk

1½ oz (40 g) grated hard cheese

salt and pepper

These quantities are sufficient to fill four 7-fl oz (200-ml) or six 5-fl oz (140-ml) timbale moulds or ramekin dishes. Preheat the oven to 350°F (180°C) mark 4. Select only the younger, firmer, crisp spinach leaves with no stalks to make up the weight; wash and blanch in a large pan of boiling salted water for 2 minutes then drain and squeeze out all excess moisture. Weigh out 3 oz (75 g) of the squeezed spinach. Peel the shallot or onion and chop very finely; sweat gently in the butter until soft, add the spinach and cook, stirring, for 2 minutes so that any remaining moisture will evaporate. Process to a fine purée in the food processor.

Grease the moulds with butter. Beat 2 whole eggs and 2 yolks lightly with a pinch of salt and a generous pinch of white pepper. Beat in the cream, milk, spinach and cheese. Pour into the moulds.

Place the moulds in a roasting tin and pour enough boiling water into it to come about two-thirds of the way up their sides. Cook in the oven for 30 minutes. Take the moulds out of the water and allow to stand for a minute, then turn out.

———— • ————

BRUSSELS SPROUTS WITH FRESH CORIANDER AND LIME JUICE

1¼–1½ lb (600 g) Brussels sprouts
2 shallots or 1 small onion
½ oz (15 g) butter
5 fl oz (150 ml) double cream
¼ beef or chicken stock cube
2 tbsp chopped coriander leaves
½ or 1 small red chilli pepper
1 lime
salt and black pepper

Heat plenty of salted water in a large saucepan. Trim and wash the Brussels sprouts but do not make the usual cross cut in their solid, stalk ends. Add to the boiling water and cook until tender but still crisp (8–15 minutes, depending on their size). Drain and cut lengthwise in half. Chop the peeled shallots or onion and sweat in the butter until soft over a very low heat in the pan used for cooking the sprouts. Add the cream, the crumbled ¼ stock cube and stir. Leave to simmer gently. Meanwhile, wash the red chilli pepper, remove the stalk and seeds and chop fairly coarsely or cut into very thin rings. Stir into the cream and then add the Brussels sprouts.

Stir very gently for 2 minutes, taking care that the sprouts do not turn mushy. Sprinkle with the lime juice, stir once more and cover the pan. Remove from the heat and leave to stand for 5 minutes, then transfer to a heated serving dish and garnish with the chopped coriander leaves.

———— • ————

ROCKET, ARTICHOKE AND EGG SALAD

8 oz (250 g) rocket or any small-leaved salad greens
8 cooked or tinned artichoke hearts
4 very fresh eggs
citronette dressing (see page 35)
1 tsp finely chopped spring onions or chives
1 tbsp olive oil
salt and pepper

Have the eggs at room temperature, place them in a saucepan and add sufficient cold water to cover them. Heat very slowly to boiling point and as soon as the water reaches a full boil, remove from the heat. Take the eggs out and leave to cool; they will be deliciously soft-boiled and should be perfectly safe to eat, provided you have bought them from a reliable source. If, however, you prefer your eggs almost hard-boiled, boil for 5 minutes after the water has reached boiling point and cool quickly in cold water.

Drain the artichoke hearts if using tinned ones, cut into julienne strips (see page 24) and place in the centre of a shallow bowl. Surround with the rocket leaves.

Peel the eggs carefully, cut in half and place next to the artichoke strips. Stir the chopped spring onion or chives into 1 tbsp olive oil with a pinch of salt and drizzle delicately over the eggs. Sprinkle the citronette dressing over the rocket leaves and artichoke strips, and serve at once.

———•———

DANDELION AND PRAWN SOUP

7 oz (200 g) very fresh young dandelion leaves

10 oz (300 g) Dublin Bay prawns

2 tbsp olive oil

2 cloves garlic

1 sprig fresh thyme or pinch dried thyme

scant 1½ pints (800 ml) shellfish stock (see page 38)

salt and pepper

Serve with:

garlic croutons (see method)

Peel the scampi or prawns if you have not already done so (to provide some of the heads and shells for making the stock) and remove the black vein or intestinal tract that runs down their backs.

Wash the dandelion leaves well and cut into thin strips. Lamb's lettuce or curly endive may be used instead of dandelion leaves, if preferred. Fry the peeled, crushed garlic gently in the oil with the thyme, then add the prawns and cook for 1 minute. Add the dandelion leaves and cook, stirring, for a further 2 minutes. Pour in the hot stock and simmer for 1 minute.

Add a little more salt if necessary and white pepper to taste. Serve at once with bread croutons fried in butter with a little crushed garlic.

LETTUCE AND PEA SOUP
Endiviensuppe (Germany)

1 head curly endive or escarole

8 oz (250 g) fresh or frozen peas

1¾ pints (1 litre) chicken stock (see page 37)

2 medium-sized waxy potatoes

1 small onion

scant 1 oz (20 g) butter

1 tbsp oil

pinch saffron threads

salt and pepper

toasted slices of French bread

Bring the chicken stock to the boil, add the peas and boil gently until very tender (about 20 minutes); put them through a mouli-légumes (vegetable mill) or allow the liquid and peas to cool a little and then process in the blender in several batches. Peel and dice the potatoes; peel the onion and chop very finely. Shred the well washed and dried curly endive leaves into a chiffonade (see page 24). Sweat the onion and endive in the butter and oil in a large saucepan until tender; add the potatoes and fry briefly, stirring. Pour in the puréed peas and stock, add salt to taste and the saffron. Simmer for 10 minutes. Season with a little freshly ground pepper and serve.

———•———

CHINESE WATERCRESS SOUP

2 bunches watercress

2 thin spring onions

4 oz (120 g) beef fillet

1 tbsp plain flour

scant 1½ pints beef stock (see page 37)

1 tbsp light soy sauce

salt and pepper

Heat the stock. Wash and drain the watercress, removing and discarding all but the smallest stalks. Cut off the roots and leaf tips of the spring onions, remove the outermost layer of the bulb and discard. Slit the spring onions (bulbs and leaves) lengthwise in half, then cut these long, thin pieces into ¾-in sections.

Place the raw fillet in the freezer for 10 minutes to make it easier to slice into wafer-thin slices. Cut each of the slices in half and then cut into fine strips; place in a bowl and sprinkle with 1 tsp salt and the flour; stir well.

When the stock has come to the boil, stir in the soy sauce, the beef strips and the watercress. Simmer gently for 1 minute only. Add the spring onion pieces and simmer for a further minute, then remove from the heat. Season with a pinch of freshly ground black pepper and serve.

———•———

CREAM OF WATERCRESS SOUP

12 oz (350 g) watercress
3 shallots
1 oz (30 g) butter
4 tbsp plain flour
scant 2 pints (1.1 litres) vegetable stock (see page 38) or chicken stock (see page 37)
2 egg yolks
5 fl oz (150 ml) double cream
salt

Wash the watercress. Drain. Take off all the leaves and their little stalks, discard all the thicker stalks. Peel the shallots or onion and chop very finely indeed. Fry in the butter for 3 minutes without browning at all. Add the watercress and a pinch of salt. Stir well, cover with a tight-fitting lid and leave to sweat over a gentle heat for 5 minutes, or until the watercress has turned dark and wilted. Sift the flour into the saucepan and stir well over the low heat for

1 minute. Take the pan off the heat and use a balloon whisk to blend in the boiling hot stock, adding this gradually to avoid any lumps forming.

Return to the heat, cover and simmer for a further 5 minutes. Turn off the heat. Use a hand-held electric beater or rotary whisk to beat the soup until it is very smooth. Just before serving the soup, reheat it to boiling point and draw aside from the heat. Beat the egg yolks with the cream in a small bowl and whisk into the soup. Place over a very low heat and continue beating for 2 minutes as the soup thickens a little. Do not allow to boil or the eggs will curdle.

———•———

RUSSIAN SPINACH AND SORREL SOUP
Šči

2¼–2½ lb (1 kg) spinach
8 oz (250 g) small sorrel leaves
1 medium-sized carrot, peeled and very finely chopped
1 large Spanish onion, peeled and very finely chopped
2 tbsp finely chopped parsley
1 oz (30 g) butter
3 tbsp plain flour
2 bay leaves
1 pint (600 ml) beef stock (see page 37)
4 hard-boiled eggs
salt and black peppercorns
For the smetana:
3½ fl oz (100 ml) whipping cream, stiffly beaten
2 fl oz (50 ml) soured cream or Greek yoghurt
1 lemon

Wash the spinach very thoroughly, using only the crisp, fresh and tender leaves and removing all but the thinnest stalks. Remove the stalks from the sorrel, wash and set aside. Sweat the spinach in a covered saucepan for 2 minutes with a pinch of salt and no added water. Place the spinach in a sieve and refresh under running cold water; squeeze out all the moisture. Purée in the food processor.

Sweat the finely chopped onion, carrot and parsley in the butter for 10 minutes over a low heat in a large covered saucepan, stirring at frequent intervals. Stir in the flour and continue stirring while cooking for about 30 seconds. Remove from the heat and pour in all the boiling stock at once; beat vigorously to prevent lumps forming. Stir in the puréed spinach, the bay leaf and a few whole black peppercorns, these last 2 tied in a small piece of muslin. Simmer gently for 10 minutes; shred the sorrel leaves into a chiffonade (see page 24), add to the soup and simmer for a further 10 minutes, stirring occasionally. Draw aside from the heat, remove the peppercorns and add salt to taste.

Slice the hard-boiled eggs thinly; make the smetana by mixing the stiffly beaten cream, the soured cream and a little lemon juice to taste. Ladle the soup into individual heated soup bowls; place 1 tbsp of the smetana in each and float the egg slices on top (one egg per serving). Serve at once.

———— • ————

JAPANESE SPINACH AND EGG SOUP

| 1 lb (450 g) fresh young spinach |
| 1 cube dashinomoto *instant fish stock* |
| 3 tbsp white wine vinegar |
| rind of ½ lemon |
| 4 very fresh eggs |
| 1 tsp Japanese soy sauce |
| salt |

Make the instant fish stock by pouring scant 1½ pints (800 ml) boiling hot water from the kettle into a saucepan containing the *dashinomoto* cube. Set 1¾ pints (1 litre) cold water mixed with the vinegar to heat in another saucepan. Select spinach leaves and

stalks that are young and tender. Wash and add to a large saucepan of boiling salted water to blanch for only 30 seconds, drain in a sieve and immediately refresh by rinsing under running cold water. Cut the lemon rind into very fine strips. When the vinegar water comes to a very gentle boil, break the eggs on to a saucer one at a time and slide them into the water at the point where the water is actually bubbling; once the whites have set, remove them carefully with a slotted spoon or ladle and place each in a heated soup bowl. Divide the spinach into 4 portions and place beside the egg in each bowl. Stir the soy sauce into the hot stock and ladle into the bowls. Sprinkle the lemon rind strips on top.

———— • ————

SWISS CHARD AND CABBAGE SOUP

| 5 oz (150 g) Swiss chard |
| 5 oz (150 g) Savoy cabbage leaves |
| generous 2 pints (1.2 litres) chicken stock (see page 37) |
| 1 small onion |
| 1 small bunch parsley |
| 1 small bunch fresh basil or small pinch dried |
| 2 tbsp extra-virgin olive oil |
| 4 oz (125 g) long-grain rice |
| scant 1 oz (20 g) unsalted butter |
| 4–5 tbsp finely grated hard cheese |
| salt and pepper |

Heat the chicken stock slowly in a large saucepan to just below boiling point. Wash the Swiss chard well once you have removed the stalks. Wash the cabbage leaves, removing the large, hard ribs, and check their weight after you have prepared them. Shred the leaves into a chiffonade (see page 24). Peel the onion and chop very finely and do likewise with the washed and dried parsley and basil. Fry the onion gently in the olive oil in a large saucepan until tender and pale golden brown. Add the shredded leaves, turn up the heat to fairly high and cook, stirring with a wooden spoon, for 2 minutes. Add 4 fl oz (120 ml) of the hot stock, cover and simmer gently for 5 minutes.

Add the rest of the hot stock, bring to the boil and sprinkle in the rice. Stir, then leave to boil gently for about 10–15 minutes. When the rice is just tender, remove from the heat, stir in the parsley, basil, butter and cheese. Season to taste and serve.

———•———

NETTLE SOUP

12 oz (350 g) nettles (net prepared weight)
1 medium-sized leek
1 oz (30 g) butter
scant 2 pints (1.1 litres) chicken stock (see page 37)
4 tbsp plain flour
5 fl oz (150 ml) single cream
salt and pepper

Use only the tender tips of the nettles, gathered when they are young shoots. When you have 12 oz (350 g) wash them well, add to a pan of boiling salted water and blanch for 2 minutes then drain and refresh under running cold water. Squeeze out as much moisture as possible and chop them very finely. Chop the leek.

Melt the butter in a large saucepan over a low heat and fry the leek gently for 5 minutes, add the nettles and cook for 2 minutes while stirring. Sift in the flour, cook for 1 minute while stirring and then pour in the hot stock while beating with a balloon whisk. Lower the heat and simmer for 30 minutes.

Allow the soup to cool a little then process to a smooth, creamy texture in the blender and return to the saucepan. Stir in the cream and season.

HERB AND RICE SOUP WITH PESTO

Approx. 7 oz (200 g) mixed herbs and wild or cultivated salad leaves (e.g. fresh borage, basil, marjoram, a little fresh sage, Swiss chard, curly endive)
7 oz (200 g) rice
5 tbsp pesto (basil sauce, see page 177)
2 tbsp extra-virgin olive oil
grated Parmesan cheese
salt

Rinse the borage and the other herbs and the leaf vegetables well, take the leaves off their stalks and cut all but the smallest into thin strips. Put them in a saucepan with 2 pints (1.2 litres) water and 1 tsp salt; cover and simmer for 30 minutes. Make the *pesto* (if fresh basil is unavailable, buy the *pesto* ready-made) and mix it with 3 tbsp of the herb broth before adding the rice. Turn up the heat and allow the herb soup to come to the boil; sprinkle in the rice, stir and boil over a moderate heat, uncovered, for 10 minutes. When the rice is just tender, stir in half the *pesto* mixture and 1 tbsp of the oil and boil for a further 2–3 minutes. Draw aside from the heat, stir in the rest of the *pesto* and oil, add salt to taste and serve with Parmesan cheese.

———•———

CREAM OF CAULIFLOWER SOUP

1 12-oz (800-g) cauliflower
1 celery stalk
1 medium-sized onion
4 parsley stalks
scant 1½ pints (800 ml) chicken stock (see page 37)
7 fl oz (200 ml) single cream
1 tsp cornflour or potato flour

2 egg yolks

2 oz (50 g) finely grated Gruyère cheese

salt and white pepper

Trim and wash the cauliflower and cut it into small pieces; you will need about 1¼ lb (600 g) net weight. Trim and prepare the celery and peel the onion; slice them both thinly. Rinse the parsley stalks and chop coarsely. Place all these vegetables in the stock in a large saucepan. Cover, bring to the boil and simmer for 25 minutes. Draw aside from the heat and beat with a hand-held electric beater until smooth or cool a little before processing briefly in the blender.

Mix the cornflour or potato flour with approx. 5 fl oz (150 ml) of the cream and stir into the soup; place over a low heat and continue stirring until it boils and thickens slightly. You can prepare the soup up to this point in advance and set aside or refrigerate until shortly before serving.

Just before serving, reheat to just below boiling point. Beat the egg yolks lightly with the remaining cream and the cheese and then beat a little at a time into the soup with a balloon whisk. Do not allow to boil. Keep beating over a gentle heat as the soup thickens slightly, take off the heat and continue beating for about 30 seconds. Add salt and pepper to taste and serve.

CURRIED CREAM OF CAULIFLOWER SOUP

Use the same ingredients and quantities as for the previous recipe, Cream of Cauliflower Soup, and add:

2 tbsp mild curry powder

1 dried chilli pepper, seeded (optional)

2 tbsp chopped coriander

1 clove garlic

Follow the Cream of Cauliflower Soup method, adding the curry powder and the crumbled chilli pepper (minus the seeds) to the chicken stock. Sprinkle the finished soup with the coriander just before serving. Rub the slices of bread with the cut surfaces of a garlic clove, cut into small squares and fry in butter in a non-stick frying pan. Curry is the perfect flavouring for cauliflower.

PICKLED CABBAGE SOUP
Sauerkrautsuppe

8 oz (250 g) tinned or bottled Sauerkraut

1¾ pints (1 litre) chicken stock (see page 37)

scant 1 oz (20 g) butter

scant 1 oz (20 g) plain flour

scant 1 tsp cumin seeds

salt and pepper

Heat the stock. Melt the butter in a large saucepan, stir in the flour and cook over a low heat for 2 minutes. Remove from the heat and pour in the stock all at once, beating energetically with a balloon whisk to prevent lumps forming. Return to the heat, add the Sauerkraut and the cumin, cover and leave to simmer gently for about 30 minutes. Add salt and pepper to taste and serve very hot.

TUSCAN BEAN SOUP
Ribollita

10 oz (300 g) kale or spring greens

1¼ lb (600 g) dried white haricot beans, pre-soaked

1 medium-sized onion

3 cloves garlic

2 leeks, trimmed
2 large tinned plum tomatoes, drained
1 carrot
1 green celery stalk
1 ham bone or knuckle
4 slices 2-day old wholemeal bread
extra-virgin olive oil
1 sprig fresh rosemary or pinch dried rosemary
1 sprig fresh thyme or small pinch dried thyme
salt and pepper

Soak the beans in cold water with 1 tsp bicarbonate of soda for 12 hours. Drain and rinse before using. Peel the garlic and onion and chop them finely; fry gently in 5 tbsp olive oil in a large fireproof casserole dish, preferably earthenware or enamelled cast iron. Prepare and clean all the other vegetables; chop the heart of the leek having discarded the tough outer layers, the carrot, the celery and the tomatoes and add to the casserole dish. Stir, cover with a tight-fitting lid and sweat over a low heat for 15 minutes, stirring at intervals. Add the beans, the ham bone and enough cold water to completely cover the contents of the casserole. Cover and simmer for 1½ hours. Add salt to taste after 1¼ hours.

Gently fry a peeled, crushed garlic clove with the rosemary and thyme in 2 tbsp oil for a few minutes, then pour the oil through a small sieve and set aside. Prepare and wash the cabbage and slice the leaves into a chiffonade (see page 24). Remove the ham bone from the casserole dish; transfer about 1½ large ladlefuls of beans to a bowl and reserve. Purée the remaining soup and transfer to the casserole dish; stir in the flavoured oil and the shredded kale or spring greens. Bring to the boil and then simmer for 1 hour or until very tender. Add salt to taste, add the reserved whole beans, stir and simmer for a further 5 minutes.

Remove from the heat and add plenty of freshly ground pepper. Place slices of stale or lightly toasted brown bread into heated individual bowls and ladle the soup on to those.

CREAM OF CABBAGE SOUP

1½ lb (700 g) Savoy cabbage
scant 1½ pints (800 ml) beef or chicken stock (see page 37)
1 clove garlic
2 sprigs parsley
2 sprigs basil
1 leek
1 medium-sized floury potato
4 oz (120 g) spicy sliceable sausage
2 fl oz (50 ml) single cream
1 oz (30 g) grated Gruyère or Emmenthal cheese
2 tbsp finely chopped parsley
salt and black peppercorns
Serve with:
croutons

Trim, prepare and wash the cabbage, cut lengthwise into quarters, take out the hard core and largest ribs, rinse and drain well. Cut the leaves into very thin strips and place in a large pot with the stock, the herbs and 5 black peppercorns. Trim and wash the leek; slice it into rings; peel the potato and cut into small pieces; add both to the soup with ½ tsp salt, cover, bring to the boil and simmer for 50 minutes. Remove the skin of the sausage (a highly flavoured and rather fatty sausage made with plenty of paprika is best) and dice. Heat a non-stick frying pan and when it is very hot, fry the sausage for 2–3 minutes, turning the pieces with a non-stick spatula. The sausage should be crisp; drain off and discard all the fat it has released and leave the pieces to finish draining on kitchen paper.

When the cabbage is very tender, use a hand-held electric beater to reduce the soup to a creamy consistency or allow to cool a little and process in batches in the blender, then return to the saucepan and reheat to just below boiling point. Draw aside from the heat; add salt to taste, stir in the cheese, cream, parsley and sausage. Sprinkle with croutons and serve.

Risotto with Chicory

7 oz (200 g) red Treviso chicory or white chicory

11 oz (320 g) risotto rice (e.g. arborio)

2 oz (60 g) butter

1 small onion

4 fl oz (120 ml) dry white wine

2¾ pints (1½ litres) beef, chicken or vegetable stock (see pages 37 and 38)

2 oz (60 g) grated Parmesan cheese

salt and pepper

Bring the stock slowly to the boil. Rinse the chicory leaves, dry in a salad spinner and shred into a chiffonade (see page 24). Peel the onion and chop very finely; sweat it in 1 oz (30 g) of the butter in a wide, fairly shallow saucepan. Add the chicory and cook gently while stirring with a wooden spoon until limp and tender.

Add the rice and cook, stirring, for a minute or two, then add the wine. Pour in about 8 fl oz (425 ml) of the boiling hot stock and leave to cook, stirring from time to time; once the rice has absorbed most of the liquid, add more. Test the rice; when it is just tender (after about 14 minutes), draw aside from the heat, and add the remaining butter and Parmesan.

———— • ————

Nettle Risotto

4 oz (120 g) tender young nettle tops (net weight)

11 oz (320 g) risotto rice (e.g. arborio)

3 large shallots or 1 small onion

2 oz (60 g) butter

4 fl oz (120 ml) dry white wine

scant 2¾ pints (1½ litres) chicken stock (see page 37)

1 oz grated Parmesan cheese

salt and pepper

Heat the stock in a large saucepan. Wash the nettle leaves. Blanch for 2 minutes in boiling salted water, drain and chop very finely. Peel the shallots, chop very finely and cook gently in half the butter in a large saucepan for a few minutes until tender. Add the chopped nettles and cook for 2–3 minutes, stirring. Sprinkle in the rice and cook over a slightly higher heat for 2 minutes while stirring. Pour in the wine.

Cook, uncovered, until all the wine has evaporated, then add about 8 fl oz (225 ml) boiling hot stock; leave the risotto to cook, stirring occasionally and adding about 4 fl oz (120 ml) boiling stock at intervals as the rice absorbs the liquid. After about 14–15 minutes' cooking time the rice will be tender but still have a little 'bite' left in it when tested; take off the heat and stir in the remaining butter which will melt and make the rice look glossy; sprinkle with the freshly grated Parmesan cheese. Add salt and pepper to taste, stirring gently.

———— • ————

Cauliflower and Rice au Gratin

1 small cauliflower

5 oz (150 g) easy-cook rice

3 oz (80 g) grated hard cheese

béchamel sauce (see page 36)

3 oz (80 g) butter

nutmeg

salt and pepper

Prepare the cauliflower, separate all the florets and cut the larger pieces of stalk into small pieces. Cook in a large saucepan of boiling salted water for 10 minutes with the bicarbonate of soda; drain well.

Place the rice in a fairly small saucepan with 12 fl oz

(350 ml) water and scant ½ tsp salt. Once the water has come to the boil, cover the pan and cook over a low heat for 15 minutes by which time the rice should be tender and should have absorbed all the water. Stir in scant 1 oz (20 g) of the butter and 3 tbsp of the grated cheese. Add a little more salt if wished. Preheat the oven to 400°F (200°C) mark 6. Make the béchamel sauce and season with salt, pepper and a pinch of nutmeg. Stir in 3 tbsp of the grated cheese.

Grease a gratin dish or any shallow, ovenproof dish with butter and spread the rice out in it evenly. Cover with a layer of the cauliflower; combine the two without breaking up the cauliflower too much and completely cover with the béchamel sauce. Sprinkle the remaining cheese over the surface and dot with the remaining butter in small flakes. Place in the oven for 15 minutes, after which the surface should be crisp and golden brown.

———— · ————

CAULIFLOWER AND MACARONI PIE

14 oz (400 g) prepared cauliflower florets

¾ lb (350 g) macaroni

3 large cloves garlic

½ dried chilli pepper or pinch cayenne pepper

6 tbsp extra-virgin olive oil

grated hard cheese

salt and pepper

Bring a large saucepan three-quarters full of salted water to the boil. Divide the cauliflower into small florets. Rinse these and drain well.

Heat 6 tbsp oil in a very wide frying pan or wok and stir-fry the crushed cloves of garlic with the cauliflower florets over a high heat for 2 minutes. Reduce the heat to low, add about 2 fl oz (50 ml) of the boiling salted water heating for the pasta, the seeded and crumbled chilli pepper or pinch of cayenne pepper, cover and simmer for about 10 minutes.

Cook the pasta in the fast boiling salted water until barely tender and still with a good deal of bite to it; drain but reserve 8 fl oz (225 ml) of the water in a measuring jug. Add half this water and all the pasta to the cauliflower, stir gently and leave to simmer uncovered until the pasta has finished cooking, adding more of the reserved water when necessary. The finished dish should be very moist. Turn off the heat. Season to taste and sprinkle grated cheese on top, if wished, mixing in well. Serve immediately.

———— · ————

PASTA WITH TURNIP TOPS

14 oz (400 g) turnip tops

14 oz (400 g) pasta shells

4 cloves garlic

1 small red chilli pepper

extra-virgin olive oil

salt and freshly ground pepper

Bring a large saucepan of salted water to the boil ready for the pasta. Trim the larger stems and ribs from the turnip tops and discard, together with any wilted leaves. (Substitute broccoli if wished.) Wash well, drain and cut into small pieces. Sprinkle the pasta into the fast boiling water to prevent it coming off the boil and cook for 14 minutes before adding your chosen green vegetable, the peeled, whole garlic cloves and the whole chilli pepper. Cook until the green vegetable is only just tender and still a little crisp (about 10–11 minutes). Reserve a little of the cooking liquid when draining the contents of the saucepan. Remove and discard the garlic and chilli pepper. Transfer to a large, heated serving dish, sprinkle with a little of the reserved liquid, the olive oil and freshly ground pepper. Serve at once, on individual heated plates.

SPINACH NOODLES WITH WALNUT SAUCE

5 oz (150 g) spinach leaves (net prepared weight)

1 lb 2 oz (500 g) strong white flour

2 eggs

4 oz (120 g) grated Parmesan cheese

salt

For the walnut sauce:

4 oz (120 g) shelled walnuts

1 large slice stale white bread

approx 4 fl oz (120 ml) milk

1 small clove garlic

1 tbsp fresh marjoram leaves

1½ tbsp toasted pine nuts

butter

3 tbsp extra-virgin olive oil

5½ fl oz (160 ml) double cream

salt

Wash the prepared spinach leaves well, drain briefly and cook for a few minutes with no added water in a saucepan with a tightly fitting lid until wilted. Leave to cool, then squeeze out a handful at a time to eliminate all excess moisture; chop very finely or process to a thick purée.

Make the pasta dough: make a well in the flour in a large mixing bowl or on a pastry board, break the eggs into it and add the spinach. Gradually mix these into the flour; knead well; the dough should be smooth and soft. Wrap in cling film and chill in the refrigerator for at least 30 minutes, then roll out on a lightly floured pastry board or working surface with a floured rolling pin into a very thin square sheet. Fold the top and bottom edges over to make 3-in (8-cm) 'hems' or turnings. Keep turning both sides over towards the centre until the 2 flattish rolls meet in the centre. Take one folded side and place it on top of the other by making a final fold, placing the two flattish rolls on top of one another. Cut firmly

down across these rolls with a very sharp, large knife into ⅛-in (3-mm) wide slices, squashing the layers together as little as possible and unravelling each slice once cut. Spread out and leave to dry for at least 1 hour; if these very thin ribbon noodles (*taglierini*) are not dried they will tend to stick to one another when cooked. If you prefer, buy ready-made pasta.

While the noodles are drying, make the walnut sauce. Soak the bread in the milk, squeeze out excess moisture and place in the food processor with the walnuts, peeled garlic clove, marjoram, a pinch of salt and the pine nuts. Process briefly, just enough to make the mixture smooth. Turn into a mixing bowl and gradually mix in the oil and cream, adding a little at a time; add a little more salt to taste. Bring a very large saucepan three-quarters full of salted water to a boil, add the noodles and keep at a fast boil for 2–3 minutes (longer if using commercially prepared noodles), or until tender with just a little bite left. Drain and stir in the walnut sauce. Serve at once with grated Parmesan cheese.

——— • ———

PASTA WITH SPINACH AND ANCHOVIES

5 oz (150 g) spinach leaves (net prepared weight)

12 oz (350 g) pasta shapes (e.g. quills or shells)

2 oz (60 g) sliced Mortadella

7 fl oz (200 ml) fairly thick béchamel sauce (see page 36)

3 small tinned anchovy fillets, finely chopped

3 ripe or tinned tomatoes

1 oz (30 g) butter

3 tbsp oil

1 clove garlic, finely chopped

1 piece fresh ginger, peeled and finely chopped

1½ oz (40 g) grated Parmesan cheese

3½ fl oz (100 ml) double cream

salt and pepper

Bring plenty of salted water to the boil in a large saucepan. Bring another saucepan of salted water to the boil and blanch the well washed spinach leaves in it; drain and leave to cool, then squeeze out all the moisture and chop very finely. Chop the Mortadella and stir it into the béchamel sauce. Chop the garlic and ginger finely. Chop the anchovy fillets separately. If using fresh tomatoes, blanch and skin them, and remove the seeds. Dice the flesh. Add the pasta to the fast boiling salted water.

While the pasta is cooking, make the sauce: cook the garlic and ginger briefly in the butter and oil; draw aside from the heat and stir in the chopped anchovy fillets, crushing these and working them into the oil with a wooden spoon. Return to a slightly higher heat, add the spinach and cook for 1 minute while stirring. Stir in the béchamel sauce, the Parmesan and the cream. Drain the pasta when just tender and add to the sauce; stir over a low heat for 1 minute. Add salt and pepper to taste, then gently stir in the diced raw tomato.

———•———

SPINACH AND CHEESE DUMPLINGS WITH SAGE BUTTER

2¾ lb (1.2 kg) spinach
2 eggs
11 oz (300 g) ricotta cheese
2 oz (60 g) grated hard cheese
2 tbsp plain flour
nutmeg
2 oz (60 g) butter
1 tsp finely chopped fresh sage leaves
salt and freshly ground black pepper

Bring a very large, wide saucepan of salted water to the boil. Trim the spinach; use only the smaller, younger leaves for this dish. You should have about

1½ lb (700 g) net weight. Wash the spinach very thoroughly; blanch for 1 minute in the boiling water. Take out of the water with a slotted ladle and drain in a colander. Reserve the blanching water. When the spinach is cool enough to handle, squeeze out all the moisture by hand and chop finely. Place in a large non-stick frying pan and cook over a low heat for about 2 minutes, stirring with a wooden spoon. Transfer to a bowl and leave to cool.

Beat 1 whole egg and 1 yolk in a mixing bowl with 1 tsp salt, a little pepper and a pinch of nutmeg. Beat the spare egg white lightly in a small bowl and set aside. Blend in both types of cheese, followed by the spinach. Sprinkle in the flour and stir very thoroughly, adding salt to taste. If the mixture is very soft, add an extra 1 tbsp flour. Cover the bowl with cling film and chill for at least 2 hours.

Bring the reserved blanching water back to the boil. Take the mixture out of the refrigerator and use a tablespoon dipped in cold water repeatedly to make slightly flattened, egg-shaped dumplings, shaping these firmly in the spoon with the help of your fingers dipped in the spare egg white to prevent the mixture sticking to your fingers. Space out the shaped dumplings on a large plate or board as you prepare them. Melt the butter gently in a small saucepan and add the sage. With the water at a very gentle boil, barely more than a simmer, carefully add the dumplings one at a time. They must have plenty of room; cook in batches if necessary. At first they will sink to the bottom; after about 3 minutes they will bob up to the surface. As they do so, remove them with a slotted spoon and drain in a colander. Handle them with care as they break up easily. Sprinkle with the sage butter and serve.

———•———

PASTA WITH BROCCOLI IN HOT SAUCE

1 lb (500 g) purple sprouting or green calabrese broccoli
10–12 oz (320–350 g) pasta shapes (e.g. shells or quills)

4 garlic cloves, crushed

1–2 chilli peppers, dried or fresh

6 tbsp extra-virgin olive oil

salt

Bring a large saucepan with plenty of salted water in it to the boil. Divide the broccoli into small florets. Rinse well and drain.

Remove the seeds from the chilli peppers if using fresh, and cut into thin rings; if using dried chillis, crumble them. Heat the oil in a large, non-stick frying-pan and cook the garlic very briefly (do not allow to colour). Add the chillis and broccoli and sprinkle with a pinch of salt. Stir-fry over a high heat for 1–2 minutes, then turn down the heat to very low, cover and cook for 10–15 minutes, adding a little water at intervals to prevent the contents catching and burning. The vegetables should be tender but still crisp. Add the pasta to the saucepan of fast boiling water and cook until just tender. Reserve about 4 fl oz (120 ml) of the cooking water, drain off the rest and add the pasta to the broccoli, together with some or all of the reserved water. Cook, uncovered, for 1 minute, using a spatula to merely lift the pasta off the bottom of the wok or pan, allowing the broccoli to mix without stirring it.

———•———

RIBBON NOODLES WITH ROCKET, TOMATO AND CHEESE SAUCE

5 oz (150 g) rocket leaves (net trimmed weight)

8 oz (250 g) fresh taglierini (thin ribbon noodles, see recipe for Spinach Noodles with Walnut Sauce on page 66)

2 medium-sized ripe tomatoes

1 small clove garlic

1½ oz (40 g) grated hard cheese

6 tbsp extra-virgin olive oil

salt and black pepper

Buy commercially prepared thin ribbon noodles (taglierini) if you prefer. Heat plenty of salted water in a large saucepan in which to cook the pasta. Make the sauce: weigh the rocket after removing the stalks, wash well and dry in a salad spinner. Pour boiling water on to the tomatoes in a bowl, wait 10 seconds then drain and fill the bowl with cold water; drain again and peel, remove the seeds and any hard stalk sections and chop the flesh coarsely. Place the rocket leaves in the food processor with the peeled garlic clove and grated cheese and process just enough to chop finely, not to turn into a paste. Turn into a mixing bowl, stir in the oil and the drained, chopped tomatoes. Add a little salt and plenty of freshly ground pepper. Add the pasta to the boiling salted water, stir and then leave to cook for 3 minutes if fresh noodles are used; commercially prepared pasta will take longer. Drain when just tender, reserving about 4 fl oz (120 ml) of the liquid. Return the pasta to the saucepan, sprinkle with some of the reserved hot water so that it is very moist and then stir in the sauce; serve at once.

———•———

STUFFED LETTUCE ROLLS

2 large round lettuces (e.g. Webb's Wonder)

7 oz (200 g) closed cap mushrooms

7 oz (200 g) very thin veal escalopes

1 clove garlic

2 oz (50 g) lean cooked ham

2 oz (50 g) Parma ham

few fresh marjoram leaves

5 tbsp grated hard cheese

1 egg yolk

1¾ pints (1 litre) beef stock (see page 37)

butter

wine vinegar

salt and pepper

Serve with:

croutons

Heat the beef stock slowly. Bring a large saucepan of salted water to the boil. Prepare the lettuce, selecting the middle, fairly large leaves: you will need 12 oz (350 g) of these. Reserve the outermost, old leaves and the very small ones for use in other dishes.

Spread out 3 clean dry cloths side by side on a working surface. Blanch the lettuce leaves one by one for 30 seconds in the saucepan of boiling salted water, remove each one with a slotted ladle and spread out flat on the cloths with the greener, inner side uppermost, taking care not to tear them.

Wash, dry and chop the mushrooms. Fry the veal escalopes lightly in a little butter with a peeled whole clove of garlic; sprinkle with a little salt and pepper when cooked, chop into fairly small pieces and set aside. Cut both types of ham into strips, then chop finely. Process the veal, ham and the marjoram leaves briefly in the food processor, just enough to chop further and mix well without turning the mixture into a paste.

Transfer this stuffing to a large mixing bowl, stir in the egg yolk, the grated cheese, a little salt and plenty of pepper. Stir until the mixture is evenly blended and homogenous.

Place 1–2 tsp of this stuffing in the centre of each lettuce leaf, fold the rib end of the leaf over the filling, fold over the two sides and then roll up, forming a neat parcel; press the edge of the leaf gently to seal. When all the leaves and stuffing have been used up, arrange the parcels in a single layer, with the joins downwards, in a shallow fireproof casserole dish; sprinkle with about 8 fl oz (225 ml) of the boiling hot stock, cover and simmer gently for 5 minutes.

Fry the bread croutons in butter until golden brown and very crisp.

Carefully arrange an equal number of the lettuce parcels side by side in 4 wide, heated, soup dishes; sprinkle some croutons on top of each serving, ladle in the remaining hot broth and serve at once.

———•———

GREEN GNOCCHI

1 lb (450 g) rocket
1 lb 2 oz (500 g) floury potatoes
5 oz (150 g) watercress leaves (net weight)
1 spring onion
1 clove garlic
1 large bunch basil
olive oil
9 oz (250 g) plain flour
3 eggs
nutmeg
salt and white pepper
For the vegetable and herb sauce:
1 small clove garlic
1 small carrot
1 small green celery stalk
1 spring onion
1 sprig parsley
1 sprig basil
1 bay leaf
1 sprig sage
1 3-in (7–8-cm) white part of leek
2 oz (50 g) rocket (net weight)
4 oz (125 g) lamb's lettuce (net weight)
2 medium-sized ripe tomatoes
extra-virgin olive oil
4 tbsp grated Parmesan cheese
salt and pepper

Boil the potatoes until tender. Blanch the rocket and watercress leaves in boiling salted water for 1 minute; drain, squeeze out the moisture and chop. Chop the spring onion and garlic with a few basil leaves; fry in 2 tbsp olive oil. Add the chopped rocket and watercress, salt and pepper and a pinch of nutmeg. Cook, stirring over a low heat for a few minutes, then turn up the heat a little to dry out. Blend to a smooth paste in the food processor.

Peel and mash the potatoes. Stir in the vegetable purée, sift in the flour and work this in by hand or with a wooden spoon, gradually blending in 1 whole egg and 2 extra yolks, salt, pepper and nutmeg. Knead well until very smooth; divide the dough up into several even portions and roll each out into a long sausage shape of even thickness, about the width of your middle finger; cut into ¾-in (2-cm) sections and press each section against the prongs of a fork. Space out the gnocchi on a clean floured cloth or board.

Make the sauce: Chop the garlic, carrot, celery, spring onion, parsley and basil together finely. Cook this mixture gently in 3 tbsp olive oil in a very large frying pan for 10 minutes, stirring frequently. Tie up the bay leaf, sage and sliced leek in a muslin bag. Wash all the leaves of the green salad vegetables, dry well and chop finely. Peel the tomato, remove the seeds and chop.

Add the chopped salad greens and tomatoes to the pan together with the muslin bag, season, cover and cook for 15 minutes. Discard the bag.

Add the gnocchi in several batches to a very large saucepan of boiling salted water; when each batch bobs up to the surface, they are cooked. Remove with a slotted ladle and place in a colander then cook the next batch. Stir the gnocchi into the hot sauce, sprinkle with Parmesan, dot with flakes of butter and serve. Serves 6.

GREEN RAVIOLI WITH WALNUT SAUCE

14 oz (400–450 g) escarole
1½ lb (700 g) mixture of Swiss chard and fresh borage and fresh chervil
2 eggs
1 small clove garlic, peeled and crushed
3 oz (90 g) grated Parmesan cheese
5 oz (150 g) ricotta cheese
14 oz (400 g) strong white flour
2 tbsp dry white wine
walnut sauce (see page 66)

Cook the escarole, Swiss chard, borage and chervil in a covered pan with no added liquid until wilted and tender, drain and leave to cool and then squeeze out all the moisture. Chop finely, transfer to a mixing bowl and combine with the beaten eggs, the garlic, 2½ oz (70 g) of the Parmesan, the ricotta and a little salt. Cover and chill in the refrigerator.

Make the pasta dough: make a well in the centre of the flour and pour in the wine and approx. 5–6 fl oz (150–180 ml) hot or cold water. Add a generous pinch of salt. Mix, adding a little more water if necessary. Knead very well until smooth and elastic. Roll out into a very thin sheet on a lightly floured board and cut into rectangles measuring approx. 2¾ × 2 in (7 × 5 cm). Place 1 heaped tsp of the chilled stuffing in the centre of each and fold over, pinching to seal (moisten if necessary).

Cook the ravioli in boiling salted water for 5 minutes. Serve with walnut sauce and the remaining Parmesan cheese.

LETTUCE AND BLACK OLIVE PIZZA

2 large escaroles
scant 1 oz (20 g) fresh baker's yeast or scant ½ oz dried yeast
pinch caster sugar
1 lb 2 oz (500 g) strong white flour
olive oil

2 cloves garlic
½–1 dried chilli pepper (optional)
15 large black olives, stoned
salt and pepper

Crumble the yeast into 3½ fl oz (100 ml) lukewarm water and add a pinch of sugar. Stir gently. Leave to stand for 10 minutes at warm room temperature or until the surface is covered with foam. Sift the flour into a large mixing bowl with 1 tsp salt, make a well in the centre and pour in the yeast mixture and 2 tbsp light olive oil.

Use your hand or a wooden spoon to stir the yeast and oil, gradually incorporating the flour. When it is all combined, transfer to a lightly floured pastry board and knead energetically for about 5 minutes or until the dough is very light, smooth and elastic. Shape into a ball and return to the mixing bowl; place a clean, damp cloth over the bowl and leave to rise in a warm place away from draughts for 30 minutes or longer, until doubled in bulk.

Heat 2 tbsp oil in a very large saucepan and gently fry the lightly crushed, whole peeled garlic cloves for a minute to flavour the oil, then remove them. Crumble the chilli pepper into the oil, add the escarole, a pinch of salt and some freshly ground pepper and cook over fairly high heat for 2 minutes, stirring continuously. Cover the pan, reduce the heat to very low and cook with no added liquid for 10 minutes, then take the lid off and stir in the olives; cook over a higher heat for a few minutes to allow all the liquid produced by the lettuce to evaporate. Add a little more salt and pepper and remove from the heat.

When the dough has doubled in bulk, preheat the oven to 475°F (240°C) mark 9. Flatten the dough with a floured rolling pin on a floured board and roll out into a disc 12 in (30 cm) in diameter.

Place in a well oiled flan tin or quiche dish of the same diameter, pressing the dough with your knuckles and spreading it to line the base and sides. Spread the lettuce and olive mixture evenly over the pizza dough, sprinkle with a very little olive oil and bake in the oven for 15 minutes.

——— • ———

APULIAN PIZZA PIE

2 large escaroles
5 fl oz (150 ml) beef stock (see page 37)
2 cloves garlic
15 firm, juicy black olives, stoned
2 tbsp capers
2 tbsp sultanas
3 tbsp pine nuts
1 lb 2 oz (500 g) strong white flour
scant 1 oz (20 g) fresh baker's yeast or scant ½ oz (10 g) dried yeast
pinch caster sugar
butter for greasing
olive oil
caster sugar
salt and pepper

Make the pizza dough as described in the previous recipe and leave it to rise. Make the filling: trim and wash the escarole, discarding the outermost leaves and separating the rest. Cook these in the stock in a large, covered saucepan over a low heat for 20 minutes by which time they will be wilted and tender. Heat 2 tbsp olive oil in a large frying pan, add the whole, peeled and partly crushed cloves of garlic and cook gently until pale golden brown, then discard the garlic. Turn up the heat a little and add the contents of the saucepan without draining off any liquid. Stir well and cook for 5 minutes. Sprinkle with a pinch of salt, add some freshly ground pepper and the olives, capers and sultanas. Stir, cover and simmer over a low heat for a further 5 minutes.

Grease a non-stick pan with a little butter and cook the pine nuts gently, stirring continuously, until they are lightly coloured. Stir into the endive mixture and take off the heat; taste for seasoning. Preheat the oven to 360°F (180°C) mark 4 when the dough has doubled in bulk. Divide the dough in half and roll out into 2 thin discs on a lightly floured board; roll out one to a diameter of 12 in (30 cm) and use to line a deep 10-in (26-cm) cake tin or pie dish. Dip a pastry brush in oil and brush a thin film of it over the surface of the dough lining. Fill with the escarole mixture, levelling out the surface and drizzle over a few drops of oil. Roll out the second disc to a diameter of about 10½ in (27 cm), and place over the pie. Press the edges together, folding them over, away from the rim, in a narrow fold all round the edge and press with the prongs of a fork to make a decorative border. Brush the surface lightly with a little oil and bake for 30 minutes, until golden brown. Serve hot or warm.

———— • ————

CHICKEN AND CHICORY GALETTE

For the aromatic chicken stock:
1 3–3¼-lb (1½-kg) free-range chicken
1 leek
2 cloves
1 celery stalk
1 baby turnip
1 carrot
few parsley stalks
1 freshly peeled tomato skin
1 sprig fresh thyme
1 bay leaf
salt and peppercorns
For the pancakes:
see recipe on page 36
For the filling:
1 head radicchio
3 heads red Treviso chicory or white chicory
1 medium-sized onion
1 oz (30 g) butter
meat from the chicken, taken off the bone
7 fl oz (200 ml) creamy velouté sauce (see below)
5 fl oz (150 ml) hot aromatic chicken stock (see above)

salt and pepper
For the creamy velouté sauce:
2 oz (60 g) butter
2 oz (60 g) plain flour
generous 1½ pints (900 ml) aromatic chicken stock (see above)
3½ fl oz (100 ml) double cream or crème fraîche
nutmeg
salt and white pepper
For the topping:
5 oz (150 g) thinly sliced Gruyère or Emmenthal cheese
5 oz (150 g) coarsely grated Parmesan cheese
2 tbsp butter

Use 2 soufflé dishes or similar deep, ovenproof receptacles about 9½–10 in (24–26 cm) in diameter for this recipe. You will need to cook the chicken, take the meat off the bone and make the pancakes the day before you plan to serve this dish. Place the chicken in a very large saucepan with 2¾ pints (1.6 litres) water, a pinch of salt, the well washed and trimmed white part of the leek stuck with the 2 cloves and all the other ingredients listed above under stock. Cover and bring slowly to the boil; remove the lid, turn up the heat a little and remove any scum as it rises to the surface. Simmer for about 30 minutes or until the chicken is cooked (test by inserting a pointed knife deep into the thigh; if the juices run clear it is done). Take the meat off the bone when cool enough to handle. Strain the stock and refrigerate overnight. Refrigerate the chicken meat.

Make the pancake batter following the recipe on page 36, cover and refrigerate for at least 6 hours before making the pancakes; use more batter for each one than usual as you will need fairly thick pancakes; make them in a frying pan of a slightly smaller diameter than the soufflé cases in which you will be layering them with the filling. This batter should yield 10 pancakes, eight of which should be good enough to use. Pile them up on top of one another, cover with cling film and refrigerate overnight. The next day remove all the fat that will have solidified in a layer on top of the chicken stock. Strain the stock again, this time through a piece of muslin placed in a sieve or through a very fine sieve. Measure out 4 fl oz (125 ml), add salt to taste and heat. Preheat the oven to 400°F (200°C) mark 6. Cut the chicken meat, without the skin, into thin strips. Trim the radicchio and Treviso chicory or ordinary chicory, rinse, drain and cut into thin strips. Peel the onion and chop very finely; cook gently in 1 oz (30 g) of the butter in a large saucepan until tender but not browned. Add the radicchio and chicory leaves, sprinkle with a pinch of salt and cook stirring, for 3–4 minutes. Pour in 3–4 fl oz (90–120 ml) of the stock, cover tightly, and cook over moderate heat for 5 minutes, until the vegetables are wilted and tender; continue cooking briefly, uncovered, to let the remaining liquid evaporate. Draw aside from the heat and stir in the chicken strips.

Make the creamy velouté sauce: melt the butter, stir in the flour and cook over a low heat for 2 minutes. Draw aside from the heat and add all the hot stock at once, beating quickly. Return to the heat and simmer for a few minutes, stirring, until the sauce is very smooth and glossy. Stir in the cream and add a little salt, freshly ground white pepper and nutmeg to taste; simmer for a further 2 minutes then take off the heat. Add about 7 fl oz (200 ml) of this sauce to the chicken mixture and fold in gently.

Grease 2 soufflé dishes about 9½–10 in (24–26 cm) in diameter with butter (trim the pancakes to fit if necessary). Cover the bottom of the ovenproof dishes with a thin layer of the remaining sauce; cover this with a pancake, laid flat. Spread a thin layer of sauce on top; divide one third of the chicken and vegetable mixture between the 2 dishes, spreading it out evenly and stopping about 1 in (3 cm) short of the edges. The next layer calls for both cheeses: divide a quarter of each cheese between the 2 dishes. Place the second pancake on top; spread another thin layer of sauce, cover with the same quantity as before of the chicken and vegetable filling, a cheese layer, followed by the third pancake. Repeat this sequence one more time; when the last 2 pancakes are in place, spread a generous layer of the remaining sauce on top and use up

the remaining quarter of the cheeses for a last, even layer. Dot with butter. Place in the oven for 15–20 minutes or until the surface has browned. Turn off the oven, leave to stand for a few minutes with the door ajar and then serve, cutting down through the layers like cake. Serves 8–10.

———— • ————

STUFFED LETTUCE

2 or 3 large Webb's Wonder lettuces

7 oz (200 g) boneless chicken breast, skinned

2 slices white bread, crusts removed

7 fl oz (200 ml) milk

1 oz (30 g) dried mushrooms, pre-soaked (e.g. ceps)

4 oz (100 g) cooked lean ham

1 clove garlic

1 tsp fresh marjoram leaves

4 oz (100 g) curd cheese

2 oz (50 g) grated Parmesan cheese

3 eggs

14 fl oz (400 ml) tinned creamed tomatoes (passato*)*

olive oil

salt and pepper

Choose firm, flavoursome lettuces for this recipe; Webb's Wonder or escarole would be suitable. Bring plenty of salted water to the boil in a large saucepan. Use the outermost lettuce leaves provided they are undamaged as well as the next couple of layers of large leaves: you will need at least 20 leaves. Save the rest of the lettuce for salads. Rinse the leaves well. Spread 3 or 4 clean cloths flat on the work surface. Blanch the leaves one at a time for 30 seconds, take out and spread flat on the cloths with their inner, greener sides facing uppermost, handling them with great care.

Fry the chicken lightly in a little oil; sprinkle with a little salt and pepper when done, slice into thin strips and set aside. Soak the bread in the milk, then squeeze out all excess moisture. Have the dried mushrooms ready soaked in a small bowl of warm water (use dried Italian ceps or *porcini* if possible as these have plenty of flavour); squeeze out all excess moisture by hand when they have reconstituted. Shred the ham and process briefly with the chicken, mushrooms, peeled garlic and marjoram to chop finely; do not process into a paste. Transfer to a large mixing bowl and work in the ricotta, Parmesan, the eggs (one at a time) and the moistened bread. Season with salt and pepper.

Place 1–2 tsp of this mixture in the middle of each lettuce leaf, adjusting the quantity to the size of the leaf. Fold the rib or stalk end of the leaf over the filling, fold the two sides over each end and roll up, enclosing the filling neatly and pressing down gently on the joins. Pour two thirds of the *passato* into a very wide, shallow pan; season lightly if wished. Place the lettuce rolls in it, in a single layer. Cover with the remaining sieved tomato. Heat gently to reach boiling point and then simmer gently for exactly 6 minutes with the lid on. Take the rolls out of the pan and place on individual plates with some of the tomato liquid; serve warm rather than very hot.

— • —

CHICORY AND CHEESE MOULD WITH TOMATO SAUCE

2¼–2½ lb (1 kg) chicory
2–3 oz/60–90 g thinly sliced Parma ham or cooked ham
2 eggs
2 large shallots, finely chopped
18 fl oz (500 ml) milk
7 fl oz (200 ml) velouté sauce (see page 35)
1 oz (30 g) grated Parmesan cheese
scant 1 oz (20 g) grated Gruyère or Emmenthal cheese
1 oz (30 g) butter
1 tbsp olive oil
fine dry breadcrumbs
salt and white pepper
For the tomato sauce:
2 cloves garlic
1 sprig thyme
18 fl oz (500 ml) tinned creamed tomatoes (passato)
2 fl oz (50 ml) dry white wine
pinch caster sugar
pinch oregano
1 tbsp olive oil
salt and pepper

Preheat the oven to 350°F (180°C) mark 4; grease a 2-pint (1.2-litre) soufflé dish lightly with oil and sprinkle in about 2 oz (50 g) breadcrumbs, turning the dish around and tipping it this way and that to coat the oiled surfaces evenly; tip out any excess.

Bring a large saucepan of salted water to the boil; prepare the chicory (see page 21) and blanch the leaves in the boiling water for 5 minutes. Drain them well. Cut the Parma ham into strips. Heat the butter in a very large non-stick frying pan and fry the chopped shallots gently for 5 minutes, stirring continuously. Add the ham and cook for a few seconds while stirring, then add chicory. Continue stirring for 1 minute, then increase the heat a little and gradually stir in the milk. Simmer, uncovered, for 20 minutes, stirring at frequent intervals. Season.

Make a thick velouté sauce following the recipe given on page 36. Allow to cool for a few seconds before beating in the egg yolks one at a time with a wooden spoon, followed by both cheeses. Turn into a large mixing bowl and combine with the chicory mixture. Taste and add seasoning if needed.

Pour boiling water from the kettle into a roasting tin and place in the oven; the water should be about 1½ in (4 cm) deep. Beat the egg whites with a small pinch of salt until very stiff but not dry or grainy; fold gently into the mixture in the mixing bowl, using a mixing spatula. Turn into the soufflé dish, tap the bottom of the dish on the work surface to make the mixture settle and level, sprinkle the surface with a fine layer of breadcrumbs and place the dish in simmering water in the roasting tin. The water should come just over half way up the side of the dish; top up with more boiling water if necessary. Bake for about 40 minutes or until a deep golden brown on top and springy in the centre.

While it is cooking, make a mildly flavoured, light tomato sauce: peel the garlic cloves, leave them whole but crush them slightly if you want a good garlic flavour and fry gently over a low heat with the thyme in 1 tbsp olive oil. Add the *passato*, the sugar, tsp salt and 2 fl oz (50 ml) dry white wine. Cover and simmer for 20 minutes. Remove the thyme and the garlic cloves, stir in 2 fl oz (50 ml) of water, add salt and pepper to taste and a pinch of oregano and cook, uncovered, for 1 minute while stirring.

Take the cooked mould out of the oven and leave to stand and settle for 5 minutes before running the sharp point of a knife round the inside edge of the dish. Turn a heated serving plate upside down on top of the soufflé dish and hold the two dishes together as you turn the serving dish the right way up again and release the mould on to it. Serve with the tomato sauce.

———•———

ESCAROLE WITH BEANS AND OLIVE OIL

1¾ lb (800 g) escarole
1 lb 2 oz (500 g) dried cannellini beans or white kidney beans
2 cloves garlic
2 sage leaves
1 tsp bicarbonate of soda
extra-virgin olive oil
salt and pepper

Soak the beans overnight in enough cold water to amply cover them mixed with the bicarbonate of soda. The next day, drain and place the beans in a measuring jug to get a rough estimate of their volume and transfer to a fireproof earthenware or enamelled cast iron casserole dish with twice their volume of water. Add the unpeeled garlic cloves, sage leaves but no salt and bring slowly to a gentle boil.

Simmer the beans over a very low heat for 2½ hours or until they are tender, removing any scum from the surface at intervals. After just over 1 hour's cooking time, stir in 2 tsp salt and 1 tbsp olive oil. When the beans are cooked, remove and discard the sage leaves; squeeze the soft purée from the garlic skins into the beans and stir. Add a little salt if necessary. Trim, prepare and thoroughly wash the escarole; blanch for 2 minutes in a large saucepan of boiling salted water. Drain well and place an equal amount in deep, heated soup bowls; ladle plenty of beans on top and sprinkle a little olive oil and freshly ground white or black pepper over them. Serves 6.

———•———

GRILLED CHICORY WITH CHEESE AND CHIVE DUMPLINGS

4 heads chicory
16 rashers smoked streaky bacon
8 oz (250 g) fresh curd cheese
1 egg yolk
1 small bunch chives
1 oz (30 g) grated Emmenthal or Edam cheese
2½ oz (70 g) butter
scant 2 fl oz (45–50 ml) extra-virgin olive oil
salt and pepper

Preheat the oven to 400°F (200°C) mark 6 and turn on the grill ready for use. Trim and rinse the chicory then cut each head lengthwise in half and place cut side uppermost in a single layer in a lightly oiled gratin dish. Season and sprinkle with the olive oil.

Place the curd cheese in a large mixing bowl and blend in the egg yolk with a wooden spoon; add about 1 tbsp chives snipped with kitchen scissors, the grated cheese, salt and pepper. Stir well and then mould into flattened oval shapes.

Grill the bacon until crisp; drain off all the fat. Place the gratin dish containing the chicory under the very hot grill for 8–10 minutes, turning the chicory pieces half way through the cooking time. Transfer the gratin dish to the preheated oven for 5 minutes. By now the chicory will have wilted and softened; fold each one in half so that the cut sides meet and place a piece of crispy bacon on top of each chicory half, then fill the spaces with the cheese and chive dumplings. Return the gratin dish to the oven for 4–5 minutes while you melt the butter in a saucepan. Pour the butter over the dumplings and serve.

———— • ————

SWISS CHARD AU GRATIN

1¼ lb (600 g) Swiss chard
2 oz (60 g) small button mushrooms
2 oz (60 g) butter
1 tbsp oil
1 large shallot, finely chopped
½ tbsp finely chopped parsley
14 fl oz (400 ml) béchamel sauce (see page 36) or velouté sauce (see page 35)
2 egg yolks
1½ oz (40 g) grated Gruyère or Emmenthal cheese
4 tbsp fine white breadcrumbs
6 tbsp freshly grated hard cheese
salt and cheese

Preheat the oven to 400°F (200°C) mark 6. Bring a large saucepan of salted water to the boil. Cut off the green leaves from the Swiss chard and discard. Cut the wide, white stalks into 2½-in (6-cm) sections. Boil in the salted water for 15–20 minutes or until tender but still with a little bite left in them, then drain. Chop the mushrooms. Heat 1 tbsp of the butter in a non-stick frying pan with 1 tbsp oil and fry the shallot very gently, add the mushrooms and cook, stirring, for 1–2 minutes. Season and add the parsley.

Make the béchamel or velouté sauce, remove from the heat and beat in the 2 egg yolks one at a time with a balloon whisk or wooden spoon, followed by the grated Gruyère or Emmenthal and the shallot and mushroom mixture. Cover the bottom of a shallow ovenproof or gratin dish with a thin layer of this sauce, place the Swiss chard stalks evenly at top in one layer, overlapping one another, and cover with the remaining sauce. Fry the fine breadcrumbs in 1 oz (30 g) butter; allow to cool before stirring in the grated hard cheese and then sprinkle over the surface. Melt the remaining 1 oz (30 g) of the butter and drizzle over the topping. Bake for 10 minutes or until golden brown.

———— • ————

FENNEL AND LETTUCE AU GRATIN

1¾ lb (800 g) fennel
2 firm lettuces
2 tbsp plain flour
2 oz (60 g) butter
14 fl oz (400 ml) milk
3 oz (80 g) grated hard cheese
1 egg yolk
breadcrumbs
nutmeg
salt and pepper

Preheat the oven to 400°F (200°C) mark 6. Cut the lettuces in quarters and blanch in boiling salted water for 5 minutes. Drain and set aside. Prepare the fennel (see page 19), cut lengthwise into thin slices and boil in salted water for 15 minutes. Drain and set aside. Make a béchamel sauce with 2 tbsp plain flour, 1 oz (30 g) butter and 14 fl oz (400 ml) boiling milk (see method on page 36); season with salt, pepper and nutmeg. Remove from the heat and beat in half the grated cheese and the egg yolk with a whisk.

Grease an oval ovenproof dish and fill the dish with alternating deep sections of the vegetables, alternating paler broad bands of fennel with the brigher green of the lettuce. Cover evenly with the sauce and sprinkle the surface with the rest of the grated cheese and with a fairly thin layer of dry breadcrumbs. Dot flakes of the remaining butter over the surface. Bake for 20 minutes then brown briefly under a very hot grill.

— • —

SPINACH, ARTICHOKE AND FENNEL PIE

1¼ lb (600 g) spinach
10 oz (300 g) artichoke hearts or bottoms (see page 18)
1 lb (500 g) fennel
7 fl oz (200 ml) chicken or beef stock (see page 37)
14 oz (400 g) cups Quark or curd cheese
4 oz (100 g) fromage frais
3 oz (80 g) grated Parmesan cheese
2 eggs
12 oz (350 g) frozen puff pastry, thawed
5 oz (150 g) medium-hard cheese, sliced
1 tbsp chopped chives
2 cloves garlic
2 fl oz (50 ml) olive oil
nutmeg
salt and pepper

For the artichoke sauce:
2 large shallots, finely chopped
4 thinly sliced artichoke hearts (see page 18)
1 sprig thyme
scant 1 oz (20 g) butter
olive oil
5 fl oz (150 ml) stock
3½ fl oz (100 ml) double cream
lemon juice
salt and pepper

Preheat the oven to 450°F (230°C) mark 8. Trim the spinach. Wash and blanch in the boiling salted water for 2 minutes. Drain; squeeze out all the moisture when cool enough to handle. Heat 2 tbsp oil in a frying pan with 1 peeled, whole clove garlic, add the spinach and fry gently for 3–4 minutes while stirring. Remove the garlic and set the spinach aside to cool. Prepare the artichoke hearts, dropping each one into a bowl of acidulated water to prevent discoloration. Cut lengthwise into quarters, then halve each quarter lengthwise. If the artichokes are not very young and tender, use the disc-shaped bottoms and cook whole. Fry very gently in 2 tbsp olive oil with a peeled whole but lightly crushed clove of garlic and a small pinch of salt for 2–3 minutes with the lid on. Drain off any liquid and remove the garlic clove. Set aside to cool. If only the bottoms are used, slice each one horizontally in half when cool.

Cut the fennel into strips and simmer these with the stock in a covered saucepan for 10 minutes until they are tender. Place the Quark or curd cheese and the *fromage frais* in a mixing bowl with the grated cheese and work in 1 egg yolk, a pinch of salt, pepper, nutmeg and the chopped chives.

Cut off two-thirds of the pastry and roll out on a lightly floured board into a circle 13–14 in (34 cm) in diameter; use this to line a prepared 10-in (26-cm) cake tin, allowing the extra pastry to overlap the edges evenly all the way round the rim. Spread a layer of half the soft cheese mixture over the bottom and cover with all the spinach. For the next layer use the sliced cheese and cover this layer with the artichoke hearts; cover these with the remaining soft cheese mixture.

Roll out the remaining pastry to make a pie lid 10 in (26 cm) in diameter and cover the pie; fold the overlap over this lid and press the edges together against the inside rim of the tin with the prongs of a fork. Lightly beat 1 egg and use to glaze the surface. Bake for 20–25 minutes or until golden brown.

Make the sauce: sweat the chopped shallot in the butter with 1 tbsp oil for 5 minutes, add the artichokes and the thyme and cook over a low heat for 5 minutes, stirring. Add the stock, cover and simmer for 15 minutes. Remove the thyme and beat with a hand-held electric beater until smooth. Stir in the cream, season and add a little lemon juice. Serves 10–12.

———— • ————

FENNEL, SPINACH AND BEEF SALAD

2 bulbs fennel (6–8 oz/180–225 g net weight)
8 oz (200–250 g) young spinach leaves (net weight)
1 lb (500 g) beef fillet
10 oz (300 g) fresh soft cheese (e.g. Quark, ricotta or fromage frais)
2½ tbsp chopped chives
salt and pepper
For the mustard-flavoured citronette dressing:
7 fl oz (200 ml) extra-virgin olive oil
juice of 1 large, juicy lemon
1½ tsp Dijon mustard
2 tbsp finely chopped parsley
salt and white pepper

Place the beef in the freezer for 10–15 minutes to make it firm enough to slice wafer-thin, or ask your butcher to do this for you. Wash the spinach leaves and dry. Prepare the fennel (see page 19). Sieve the soft fresh cheese unless it is already very

smooth and easy to spread; beat with a generous pinch of salt, plenty of freshly ground pepper and the chives, using a wooden spoon. Spread a beef slice flat on the chopping board and spread 1 tbsp cheese mixture evenly over it, taking care not to tear the very thin meat; place another slice on top and roll up. Repeat this until you have used up all the filling and slices, placing them on an oven sheet as you prepare them. Put the beef rolls in the freezer for 10 minutes, to prepare for the next stage in preparation. Make the mustard dressing: mix the mustard, salt and freshly ground pepper with the strained lemon juice and then beat in the olive oil a little at a time. Stir in the parsley.

Slice the fennel lengthwise into very thin slices and place in a bowl; mix these with about half the dressing. Cut the very cold, firm beef rolls with a heavy kitchen knife into slices about ½ in (1 cm) thick. Toss the spinach leaves in a little of the dressing in a separate bowl to coat them all over. Place the fennel slices in the centre of a serving platter, surround them with the spinach leaves and place the beef roll slices in a circle around the edge of the platter. Sprinkle the beef roll slices with a little olive oil and freshly ground white pepper, and serve at once. Serves 8.

———— • ————

SPINACH SOUFFLÉ

8 oz (200–225 g) prepared spinach leaves (net weight)
1 large shallot
2 oz (50 g) butter
2 oz (60 g) cooked ham, very finely chopped
9 fl oz (250 ml) milk
scant 2 oz (45 g) plain flour
5 eggs
1 tsp cream of tartar
2 oz (60 g) grated Emmenthal or Gruyère cheese
2 tbsp fine dry breadcrumbs or grated Parmesan cheese

nutmeg

salt and pepper

Serve with:

mousseline sauce (see page 107)

Have the mousseline sauce ready prepared (See method on page 107). Preheat the oven to 400°F (200°C) mark 6. Wash the spinach thoroughly, drain briefly and cook in a covered pan with no added water for a few minutes until wilted and tender. Drain off any liquid and when cool enough to handle squeeze tightly to eliminate all excess moisture. Chop very finely.

Fry the finely chopped shallot in ½ oz (15 g) butter, stirring, for 2–3 minutes then add the ham and fry for a further 2–3 minutes. Add the chopped spinach, sprinkle with a small pinch of salt and cook, stirring, until all the excess moisture has evaporated. Remove from the heat.

Make the white sauce basis for the soufflé: heat the milk slowly; melt 1 oz (30 g) butter in a fairly small saucepan, stir in the flour and cook, continuing to stir, for 2–3 minutes over a low heat. Draw aside from the heat, add all the boiling milk at once, beating vigorously as you do so with a balloon whisk or hand-held electric beater to prevent lumps forming. Return to a slightly higher heat and keep stirring as the mixture comes to a gentle boil; simmer and stir until very thick.

Take off the heat, stir in a little salt and freshly ground white pepper and a pinch of nutmeg to taste. Separate the egg whites carefully from the yolks, lightly beat the latter in a small bowl and beat a little at a time into the hot sauce, then stir in the spinach mixture. You can prepare the soufflé up to this point several hours in advance but it is slightly easier to fold into the egg whites when freshly made and still warm. Cover the sauce to prevent a skin forming.

Before you are ready to proceed to the next stage of the soufflé, have the mousseline sauce ready and keep hot over simmering water; preheat the oven. Beat the 4 egg whites with 1 extra egg white until just frothy; add the cream of tartar; this will stabilize the protein and keep the egg whites stiff for longer.

Beat until very stiff but do not overbeat (when they would look dry or 'grainy'). Fold about a quarter of the egg whites into the spinach mixture, together with all but 1 tbsp of the finely grated Gruyère or Emmenthal. Fold this mixture gently but thoroughly into the remaining egg whites.

Grease a 2-pint (1.2-litre) soufflé dish with butter, sprinkle with the grated Parmesan cheese or very fine dry breadcrumbs and pour the mixture into it. Tap the bottom on the work surface to settle and level the surface, sprinkle with the remaining cheese, place in the oven, close the oven door and immediately turn down the temperature dial to 375°F (190°C) mark 5 and bake for 25–30 minutes or until the soufflé has risen up well in the dish and is a good deep golden brown on top. Serve at once with the mousseline sauce.

———— • ————

SPINACH STIR-FRY WITH PORK, EGGS AND GINGER

8 oz (200–225 g) prepared spinach (net weight)

4 eggs

4 oz (120 g) thinly sliced pork fillet

1½ tbsp cornflour

1½ tbsp Chinese rice wine, saké or dry sherry

2 oz (60 g) dried Chinese tree ear mushrooms, pre-soaked

2 oz (50 g) tinned bamboo shoots, drained

1 ¼–½-in (1-cm) piece fresh ginger

1 small leek, white part only

1½ tbsp soy sauce

1 tbsp caster sugar

2 tbsp stock

6 tbsp sunflower oil

wine vinegar or cider vinegar

monosodium glutamate (optional)

pinch baking powder

salt and pepper

Serve with:

steamed rice (see page 126)

Make the marinade: beat 1 egg lightly and pour half of it into a bowl. Add a pinch each of salt, pepper, baking powder, monosodium glutamate (if used), 1 tbsp cornflower and 1 tsp rice wine or dry sherry. Cut the pork into thin strips, add to the marinade and mix. Leave to stand for 10 minutes.

Meanwhile, wash the spinach leaves and slice them lengthwise in half, along the central rib. Drain and rinse the mushrooms then trim off any tough parts. Cut the mushrooms and bamboo shoots into thin strips. Peel the ginger and chop finely together with the leek. Beat the 3 remaining eggs lightly with the reserved half egg, seasoning with a little salt and monosodium glutamate if wished.

Make the sauce: mix the remaining cornflour with 1 tbsp cold water in a bowl and add 1½ tbsp soy sauce, 1 tbsp rice wine or dry sherry, 1 tbsp vinegar, 1½ tbsp caster sugar, 2 tbsp stock and a pinch of monosodium glutamate (optional).

Twenty minutes before you intend serving the dish, steam the rice (see page 126 for method) and keep hot while you cook the meat.

Heat 5 tbsp sunflower oil in a wok or large frying pan. Heat a large serving plate in the oven. When the oil is very hot, add the pork and bamboo shoots and stir-fry over a high heat for 1½ minutes only, keeping the strips of meat as separate as possible. Transfer to a heated serving plate. Add 1 tbsp fresh oil to the wok and when hot pour in the beaten eggs and cook briefly until they have set. Remove with a slotted spoon, draining well, and keep hot on the serving plate. Fry the leek and ginger for 30 seconds in the remaining oil. Add the spinach and stir-fry for a further 30 seconds; add the well mixed sauce, stir to combine with spinach for 30 seconds as it thickens and comes to the boil. Return all the contents of the serving plate to the wok, stir very briefly and then return to the serving plate and serve with the rice.

———•———

GREEK SPINACH AND FETA CHEESE PIE
Spanakópita

1¼ lb (600 g) trimmed spinach leaves (net weight)

2 medium-sized onions

9 oz (250 g) feta cheese

3 tbsp chopped fresh dill or 1½ tsp dried dill

3 tbsp chopped fresh parsley

5 eggs

5 tbsp double cream

2 packets puff pastry, each weighing 9 oz, thawed

4 tbsp olive oil

salt and pepper

Preheat the oven to 400°F (200°C) mark 6. Wash the spinach, drain briefly and cook in a covered pan with no extra water for a few minutes, until wilted and tender. When cool enough to handle squeeze out all excess moisture and chop very finely. Peel and finely chop the onions; finely chop or crumble the feta cheese. Beat 4 eggs lightly in a large mixing bowl with a pinch of salt and some freshly ground pepper.

Heat 4 tbsp oil in a saucepan and fry the onion gently until soft. Add the chopped spinach and a pinch of salt; stir well, cover and cook gently for 5 minutes. Stir in the dill and the parsley and cook for 2–3 minutes, uncovered, while stirring to allow any moisture to evaporate. Remove from the heat, stir in the cream and leave to cool a little before stirring in the eggs and cheese, adding more seasoning to taste.

Roll out each pastry piece on a lightly floured board into a thin sheet large enough to line a large greased baking tray. If you prefer to use filo pastry, place half the sheets from the packet one on top of the other to line the baking tray and brush each one with melted butter or oil; do likewise with the remaining sheets for the pie lid. Spread the spinach and cheese mixture over the pastry case and place the other half of the pastry on top, pinching the edges firmly together all the way round, or pressing with the prongs of a fork to seal tightly. Beat the remaining egg lightly in a small bowl and brush all over the surface of the pastry to glaze.

Bake for about 20 minutes or until the pastry has puffed up and is crisp and golden brown. Serves 8.

———— • ————

INDIAN SPICED CABBAGE AND POTATOES
Masala Gobhi

2¼–2½ lb (1 kg) Savoy cabbage

4 waxy potatoes

3 onions

1–3 green chilli peppers

1 1-in (2½-cm) piece fresh ginger

1½ tbsp coriander seeds

2 tbsp cumin seeds

2 dried chilli peppers

1 tsp dried green mango powder (amchur) *(optional)*

3 tbsp chopped coriander leaves

5 tbsp ghee (see page 36)

salt

For the chapatis:

9 oz (250 g) wholemeal chapati flour

3½ oz (100 g) plain flour

1 tbsp ghee (see page 36)

Many shops sell ready-made fresh chapatis and nan bread, both of which are suitable for this dish. Reheat chapatis at 425°F (220°C) mark 7 wrapped in foil for 10 minutes, the nan for no more than 5 minutes and brush it with the ghee as soon as you take it out of the oven. Packets of chapati mix are also available; follow the manufacturer's instructions.

Prepare the cabbage (see page 20); wash and shred the leaves. Peel the potatoes and onions and chop both coarsely. Trim the green chilli pepper, remove the stalk and seeds and chop finely (use half a chilli if you prefer a milder tasting dish). Peel the ginger and chop very finely.

Grind 1 tbsp of the coriander seeds, scant 1 tbsp of the cumin seeds and the dried chilli peppers (the latter can be seeded, if preferred, as they are very hot) very finely in an electric grinder. Put the freshly ground spices through a sieve into a bowl and mix with the mango powder if used. Heat 2 tbsp ghee in a large saucepan and fry the remaining whole coriander seeds and cumin seeds gently for a few seconds, stirring with a wooden spoon.

Add the chopped green chilli peppers and ginger and fry, stirring, for 30 seconds. Add the ground spices, cook for a few seconds and then add the prepared vegetables. Stir well, turn down the heat to very low, cover and sweat for about 25 minutes or until the vegetables are tender but keep their shape and are not at all mushy. Moisten with 1–2 tbsp water or stock at intervals. When cooked, add a little salt and sprinkle with the chopped coriander leaves. Make the chapatis. Sift both types of flour into a large mixing bowl; make a well in the centre and pour the ghee and 6 fl oz (170 ml) lukewarm water into it. With your hand cupped and fingers held together, stir the butter and water, gradually incorporating all the flour. Knead the resulting soft dough for 6 minutes. Cover with a very damp cloth and leave to stand for 30 minutes.

Knead again and divide into 15 balls of equal size. Replace in the bowl and cover again with the dampened cloth. Grease a griddle or a large cast iron pan with a little ghee if wished and heat until very hot. Take a ball of dough out of the bowl, flatten with your palms and roll out on a lightly floured board to a 5½-in (14-cm) flat round, shake off excess flour and fry for 1 minute then turn and fry the other side for ½ minute. If the pan is hot enough, the chapati will puff up; do not worry if this does not happen. Pile up the chapatis as you cook them, keeping them wrapped in a cloth.

Serve immediately with the spiced vegetable mixture. Serves 6.

CABBAGE AND HAM ROLLS

1 large Savoy cabbage
10 oz (300 g) finely chopped lean roast pork
4 oz (100–120 g) chopped cooked ham
3 oz (80 g) Italian coppa or Westphalian ham, finely chopped
2 oz (50 g) grated hard cheese
1 egg
1 clove garlic, finely crushed
2 tbsp finely chopped parsley
2–3 tbsp béchamel sauce (see page 36)
1 small onion, finely chopped
1½ oz (40 g) streaky bacon, finely chopped
nutmeg
3 sage leaves
3½ fl oz (100 ml) dry white wine
butter
salt and pepper

Select the largest undamaged cabbage leaves, wash them and blanch one at a time in a large saucepan of boiling salted water for 2 minutes; spread out with the greener, inner side facing downwards on a chopping board and use a small, sharp pointed knife to cut out the centre rib neatly. Prepare 30 leaves in this way.

Mix all the ingredients for the stuffing together in a large mixing bowl, adding them in the order listed above, finishing up with the béchamel sauce. The mixture should be moist but firm enough to hold its shape. Season with salt, pepper and nutmeg to taste. Mould by hand into oblong rissole shapes and wrap these in the cabbage leaves, placing the stuffing on the greener, smoother side of the leaves and rolling up into neat parcels, closed at both ends. You may need more than one leaf to completely enclose the stuffing; secure with wooden cocktail sticks.

Fry the onion, bacon and sage leaves in the butter in a wide non-stick frying pan until the onion is soft; add the cabbage rolls in a single layer and fry over a moderate heat for a few minutes. They should be very lightly browned on both sides. Sprinkle the wine all over them, cover tightly and turn down the heat to very low. Simmer for 25–30 minutes. Serve with creamed potatoes.

—— • ——

CAULIFLOWER FU-YUNG

1 lb (500 g) cauliflower florets
4 oz (120 g) minced raw chicken breast
1 tbsp plain flour
pinch monosodium glutamate (optional)
2 egg whites
5 fl oz (150 ml) chicken stock (see page 37)
sunflower oil
salt and pepper
2 thin slices cooked ham cut into strips
green part of 2 spring onions cut into strips
Serve with:
steamed rice (see page 126)

Bring a large saucepan of water to the boil, add a generous pinch of salt and blanch the cauliflower florets for 5 minutes. Drain and leave to cool. Mix the chicken in a bowl with scant 1 tsp salt, the flour, monosodium glutamate if used, the egg whites and the stock. Heat 3 tbsp oil in a wok or large non-stick frying pan and stir-fry the cauliflower florets over a high heat for 2 minutes. Sprinkle them with a pinch of salt, take them out of the wok with a slotted ladle and keep hot.

Add the chicken mixture to the remaining hot oil and keep stirring over a lower heat until it thickens and becomes smooth and creamy. Return the cauliflower to the wok, reduce the heat further and stir for 1–2 minutes. Transfer to a heated serving dish and garnish with the ham and spring onion strips. Serve at once with the steamed rice.

CAULIFLOWER POLONAIS

1 medium-sized cauliflower

2 tbsp finely chopped parsley

4 slices cooked ham

4 hard-boiled eggs

1 lemon

2 oz (60 g) butter

4 tbsp fine breadcrumbs

5 fl oz (150 ml) soured cream

salt and pepper

Fill a deep saucepan wide enough to take the whole cauliflower half full of water; add salt and bring to the boil. Prepare the cauliflower, trimming off the leaves and the hard remains of the stalk, and cut a deep cross in the base. Rinse the cauliflower thoroughly and place in the boiling water which should completely cover it. (Add a pinch of bicarbonate of soda or a cube of white bread to help prevent a strong smell as it cooks, if wished.) Cover and boil gently for 17–20 minutes or until tender but still firm.

Meanwhile, make the sauce: chop the parsley, the ham and the hard-boiled eggs, keeping them separate. Squeeze the juice out of the lemon and strain. Set aside. Melt the butter in a small saucepan, add the breadcrumbs and fry for 1–2 minutes until lightly browned. Stir in the lemon juice and the cream, and remove from the heat.

Use two large slotted spoons or a spoon and a slotted ladle to take the whole cauliflower out of the water in the pan; drain as you hold it over the pan and place carefully in a heated serving dish. Make sure the sauce is hot but not boiling, stir in the ham, eggs and parsley, a pinch of salt and plenty of freshly ground pepper. Pour this all over the cauliflower and serve at once on heated plates.

— • —

CAULIFLOWER TIMBALE

1 small cauliflower

3 oz (90 g) fine fresh white breadcrumbs

pinch bicarbonate of soda

1 medium-sized onion

5 eggs

2 oz (60 g) grated Gruyère or Emmenthal cheese

9 fl oz (240 ml) milk

2½ oz (75 g) butter

nutmeg

salt and white pepper

Serve with:

fresh tomato sauce (see page 258)

Preheat the oven to 325°F (170°C) mark 3. Lightly oil the inside of a 2¼–2½-pint (1.4-litre) soufflé dish and sprinkle all over with the breadcrumbs, tipping out any excess. Bring a large saucepan of salted water to the boil, adding a pinch of bicarbonate of soda or a cube of white bread to prevent the usual rather unpleasant cooking smell.

Prepare and wash the cauliflower, divide it into florets and the lower stalks into fairly small pieces. Boil for about 13 minutes or until tender; drain well. You well need just over 1 lb (500 g) cooked cauliflower for this dish; place it in a large mixing bowl and crush finely with a potato masher or fork.

Peel and chop the onion finely and cook for about 10 minutes or until soft but not browned in ½ oz (15 g) butter, stirring frequently with a wooden spoon. Scrape this out of the pan into another mixing bowl and season with salt, freshly ground white pepper and nutmeg. Break the eggs into the bowl and beat lightly. Stir in the cheese and 2 oz (60 g) of the breadcrumbs. Bring the milk to the boil with 2 oz (60 g) of the butter and then add in a thin stream to the egg mixture, beating continuously. Stir in the cauliflower and correct the seasoning. The timbale can be prepared 2–3 hours in advance up to this point and left at cool room temperature, then placed

83

in preheated oven to cook for 35–45 minutes before serving. Fill the soufflé dish with the mixture, sprinkle the remaining breadcrumbs on the surface and bake for 35–40 minutes or until a skewer pushed deep into the timbale comes out clean and dry. Leave to stand and settle for 10 minutes before unmoulding on to a heated serving plate. Pour a little of the sauce over the timbale (substitute a cheese sauce if wished).

This is delicious served with peas mixed with small pieces of cooked ham or bacon. Serves 6.

———— • ————

BROCCOLI AND PRAWN TERRINE

10 oz (300 g) calabrese broccoli

1½ lb (700 g) Mediterranean or tiger prawns

5 fl oz (150 ml) shellfish stock (see page 38)

1¾ lb (800 g) Dover or lemon sole fillets

3 eggs

1 pint (600 ml) single cream

pinch cayenne pepper or chilli powder

lemon juice

salt and pepper

Peel the prawns, remove the black intestinal tract running down their backs and simmer them for 2 minutes in a covered pan with just enough shellfish stock to cover them. Drain and set aside.

Trim the broccoli spears and cook in plenty of boiling, salted water until tender but still crisp; drain in a colander and immediately refresh under running cold water. Drain again and leave to cool. Preheat the oven to 300°F (150°C) mark 2. Cut the sole fillets into fairly small pieces and purée in the food processor. Place 2 egg yolks and 3 egg whites in the blender and process on maximum speed for 1 or 2 minutes until they become very frothy; add the cream a little at a time, processing after each ad-

dition. Add salt, pepper and the cayenne or chilli powder. Process again. Pour half the contents of the blender into a bowl. Replace the blender container, add half the puréed sole, and process briefly. Add salt, pepper and a little lemon juice. Process once more. Empty the contents of the blender container into a bowl, and repeat the operation with the reserved egg mixture and the remaining sole purée (if you have a very large capacity blender you can process the entire quantity in a single batch).

With all the blended mixture in one large bowl, taste and add more seasoning if necessary; cover with wrap or foil and chill for 15 minutes.

You will need a 4½-pint (2½-litre) oval or rectangular terrine dish, (or 2 terrines of half this capacity). Spoon the chilled mixture into a piping bag fitted with a large-gauge nozzle, and pipe the mixture neatly so that it lines the terrine (base and sides). Press the prawns into the lined sides, tails uppermost, in a well spaced line. Pipe in more fish and egg mixture, then place the broccoli spears upside down in the terrine; pipe the remaining mixture all around the broccoli and fill the terrine completely; smooth the surface with a palette knife. Place the terrine in a roasting pan and pour sufficient boiling water into the pan to come about half way up the sides of the terrine; place in the oven and cook for 40 minutes.

Unmould and serve hot. A velouté sauce, made with the shellfish stock or a homemade tomato sauce, would go well with this terrine. Serves 8–10.

———— • ————

STEAMED BROCCOLI WITH SAUCE MALTAISE

1¾ lb (800 g) purple sprouting or calabrese broccoli

For the sauce:

7 oz (200 g) butter

3 egg yolks

5 tbsp orange juice

1 tbsp lemon juice

finely grated rind of 1 orange

salt

Trim off the larger, tougher ends of the sprouting broccoli; leave the tender leaves on the stalks. Sprinkle with a pinch of salt and steam for 12–15 minutes. (Microwave them if preferred). The sauce should be made just before serving or it may separate. Have 1 oz (30 g) of the butter cut into small cubes in the ice-making compartment of the refrigerator. Melt the rest of the butter gently in a small saucepan; beat the egg yolks with a balloon whisk or hand-held eletric beater in the top of a double boiler or in a heatproof bowl until light and creamy. Beat in 1 tbsp each of lemon and orange juice and a pinch of salt. Add half the chilled cubes of butter and place over gently simmering water; beat continuously as the butter and egg yolks gradually heat and combine. As soon as the mixture starts to thicken appreciably, remove from the heat and add the remaining chilled butter, still beating continuously; when completely blended beat in the melted butter. Start by adding just a few drops at a time and then trickle it into the sauce while beating in a very thin stream. The sauce should increase in volume and thicken. Gradually stir in the remaining orange juice and grated rind. Serve the broccoli coated with the hot sauce.

———•———

BRUSSELS SPROUTS WITH HOLLANDAISE SAUCE

1 lb (500 g) Brussels sprouts
For the hollandaise sauce:
3½ fl oz (100 ml) wine vinegar
1 shallot, peeled and finely chopped
1 sprig fresh thyme or ½ bay leaf
7½ oz (210 g) unsalted butter
3 egg yolks
lemon juice
salt and white peppercorns

Prepare the Brussels sprouts, sprinkle with a small pinch of salt and steam or microwave until just tender.

Make the hollandaise sauce: place the vinegar in a small saucepan with the chopped shallot (or ¼ mild onion), 4 peppercorns and the thyme or bay leaf. Simmer gently, uncovered, to reduce to about 2 tbsp. Strain and set aside.

Have 1 oz (30 g) of the butter cut into small cubes and chilled in the ice-making compartment of the refrigerator. Beat the egg yolks with a balloon whisk or hand-held electric beater until creamy in the top of a double boiler or in a heatproof bowl that will fit snugly over a saucepan. Gradually beat in 1 tbsp iced water and a pinch of salt. Add half the chilled butter pieces and set the double boiler top or bowl over gently simmering water; beating continuously as the mixture gradually heats and the butter melts and combines with the egg yolks, adding the reduced vinegar a little at a time as you beat. When this mixture starts to thicken noticeably, remove from the heat immediately, add the remaining chilled butter and continue beating. Once this has melted and combined with the egg mixture, beat in the melted butter, starting by adding a few drops and then gradually increasing this to a thin trickle. The sauce should become light and thick. Add a little more salt and freshly ground white pepper if wished and then stir in lemon juice to taste.

———•———

BRAISED LETTUCE

2 large firm lettuces (e.g. Webb's Wonder or similar)
2 large spring onions
1 tsp cornflour
3½ fl oz (100 ml) milk
1 vegetable or chicken stock cube
3 tbsp double cream
black pepper

Remove the older or damaged outermost leaves of the lettuces, wash and cut lengthwise in quarters. Remove the roots, leaves and outer layers of the spring onions and slice the bulbs into thin rings. Mix the cornflour with the cold milk.

Place the lettuce and milk in a shallow saucepan, crumble in the stock cube, cover and simmer gently for about 20 minutes over a low heat or until very tender and wilted. Stir in the cream and remove from the heat. Sprinkle with black pepper.

———•———

STIR-FRIED LETTUCE WITH OYSTER SAUCE

2 very fresh Cos lettuces

3 tbsp sunflower oil

For the oyster sauce:

3 tbsp commercially prepared Chinese oyster sauce

1½ tbsp light soy sauce

2 tbsp Chinese rice wine, saké or dry sherry

4 tbsp cold light stock

½ tsp cornflour or potato flour

pinch monosodium glutamate (optional)

Bring a large saucepan two-thirds full of salted water to the boil. Remove any damaged outer leaves from the lettuces, cut each whole lettuce lengthwise into quarters, wash and drain thoroughly, then cut each quarter across, in half. Add to the boiling water and blanch for only 3–4 seconds, removing the lettuce pieces quickly with a slotted handle (or use a frying basket). Drain well. Prepare the sauce mixture by stirring the ingredients together in a small bowl.

Heat the oil in a wok or large non-stick frying pan; when it is very hot, add all the lettuce at once and stir-fry over a high heat for a maximum of 2 minutes. The lettuce should still be fairly crunchy. Transfer the lettuce to a serving dish, reduce the heat, stir the

sauce ingredients once more and pour into the wok, stirring. When the sauce has boiled and thickened, sprinkle all over the lettuce and serve immediately.

———•———

FRIED ENDIVE WITH GARLIC AND CHILLI

2 lb (1 kg) curly endive

3 tbsp extra-virgin olive oil

2 cloves garlic

½–1 small dried chilli pepper, seeded

salt

Bring a large saucepan of lightly salted water to the boil. Prepare the curly endive: take all the leaves off the hard remains of the stem, discard the old, outer ones, and wash very thoroughly. Blanch the leaves in the boiling water for 2 minutes, drain well, wrap in a large cloth and squeeze out excess moisture. Heat the oil in a large, non-stick frying pan, add the peeled, lightly crushed garlic cloves and the crumbled chilli pepper. Fry gently for 30 seconds, then add the endive and cook for 5 minutes.

———•———

CHICORY WITH LEMON FLEMISH STYLE

8 heads chicory

2 oz (60 g) butter

10 fl oz (300 ml) chicken stock (see page 37)

2 tsp soft brown cane sugar

2 lemons

salt and white pepper

Prepare the chicory (see page 21), cutting out part of the bitter base. Rinse, dry thoroughly and cut lengthwise in half (if the heads are very fat, cut lengthwise into quarters). Divide the butter in half between 2 frying pans or very wide, shallow saucepans with tightly fitting lids and heat until it foams. Add the chicory pieces, spreading them out in a single layer in the hot butter, and fry over a moderately high heat until lightly browned. Turn carefully and brown the other side.

Add half the stock to each cooking receptacle, cover tightly and reduce the heat to low. Simmer for 5 minutes. Mix the sugar with the strained juice of the 2 lemons, remove the lids and sprinkle the sweetened lemon juice all over the chicory. Turn up the heat a little to reduce the lemon juice and liquid and caramelize slightly, loosening the chicory gently with a non-stick spatula if they show signs of sticking. Sprinkle with a little salt and freshly ground white pepper and serve.

———•———

ESCAROLE WITH OLIVES

4½ lb (2 kg) escarole

4 cloves garlic

1 stock cube

24 large black Greek olives, stoned

4 tbsp extra-virgin olive oil

salt and black peppercorns

Prepare the escarole: wash the leaves very thoroughly, do not shake off the water clinging to them, place straight in a very large saucepan containing 18 fl oz (500 ml) water, 8 black peppercorns and the crumbled stock cube. If there are too many leaves to fit in, add them as soon as the first batch has wilted and made room for them. Cover tightly and simmer gently for 50 minutes once the water has come to the boil, adding any remaining leaves as soon as possible. Stir at intervals, turning the leaves.

When this cooking time is up, add the oil, the olives and a little salt. Cover again and continue cooking for a further 10 minutes.

Add a little more salt if wished, stir and cook uncovered for a final 15 minutes.

———•———

BRAISED CHICORY

6 heads red Treviso chicory or white chicory

2 tbsp butter

salt and pepper

Prepare the chicory as usual (see page 21), rinse and drain. Cut lengthwise in quarters. Melt the butter in a wide non-stick saucepan with a tight-fitting lid. Arrange the chicory in a single layer and sprinkle with a little salt. Cover tightly, turn the heat down very low and leave to cook for 10 minutes. Turn the chicory, cover again and leave to cook for a further 10 minutes.

———•———

ASPARAGUS CHICORY AND BROAD BEANS

2–2¼ lb (1 kg) asparagus chicory

14 oz (400 g) young tender broad beans

extra-virgin olive oil

salt and pepper

Bring a large pan two-thirds full of salted water to the boil. Prepare the asparagus chicory, using only the leaves and the younger, tender stalks. Cut into 2-in (5-cm) sections, wash and drain. Add to the boiling water and cook for 15 minutes.

Add the shelled broad beans and cook for a further 15 minutes, or until tender. Drain the vegetables. Sprinkle with olive oil and freshly ground pepper.

BUTTERED SPINACH INDIAN STYLE

2¾ lb (1.2 kg) spinach

2 medium-sized onions

1 tsp finely chopped fresh ginger

1–2 small green chilli peppers

5 tbsp ghee (clarified butter, see page 36)

scant 1 tsp cane sugar

3½ fl oz (100 ml) garam masala

salt

Serve with:

pilau rice

Trim the spinach, removing the stalks and central ribs; wash the leaves very thoroughly. Drain in a salad spinner and shred into a chiffonade (see page 24). Peel the onions and chop very finely; trim the stalks from the chilli pepper, remove the seeds and chop the flesh finely. Fry all these chopped ingredients gently in the ghee for 5 minutes in a large saucepan, then add the spinach, a pinch of salt and the sugar. Cook over a moderate heat, stirring continuously for 2–3 minutes. Pour in the stock, stir briefly, cover and simmer for 5 minutes. Take off the lid and cook while stirring for a few seconds to reduce excess liquid. Draw aside from the heat, sprinkle with the garam masala and serve with pilau rice.

———•———

SPINACH WITH SHALLOTS

2¾ lb spinach

4 shallots

1½ oz (40 g) butter

salt and pepper

Use only the young, tender spinach leaves for this recipe; the larger, older leaves and stalks can be reserved for use in other dishes (e.g. soups, etc.). Wash and do not bother to drain them thoroughly. Peel and finely chop the shallots and cook in the butter until tender but not browned. Add the spinach leaves, fry them for a few seconds stirring to coat them with butter, then cover the pan and cook over a low heat for 2–3 minutes. They should still be fairly crisp. Sprinkle with a little salt and freshly ground pepper and serve.

———•———

SPINACH AND RICE INDIAN STYLE

2¾ lb (1.2 kg) spinach

11 oz (320 g) Patna rice

2 medium-sized onions

6 tbsp ghee (clarified butter, see page 36)

1 tsp garam masala

salt

Bring plenty of water to the boil in a large saucepan and add a generous pinch of salt. Use only the tenderest leaves, and blanch for just under 1 minute then drain; you should aim to have 1–1¼ lb (600 g) net weight of washed leaves. (The other leaves and stalks can be used for recipes calling for longer cooking and need not be wasted).

While the spinach is cooling, put the rice in a sieve and hold it under running cold water to rinse until the water runs clear. Drain. Squeeze all excess moisture out of the spinach by hand and chop finely. Peel the onions and chop very finely. Fry the onions gently in the ghee for 5 minutes, stirring with a wooden spoon until tender and pale golden brown. Add the spinach and the garam masala and stir-fry over a moderate heat for 2–3 minutes. Take off the heat, add the rice with 18 fl oz (500 ml) cold water and 1 tsp salt. Stir, return to a higher heat and bring to the boil. Cover, turning the heat down extremely low and leave to cook undisturbed for 15 minutes or until the rice is tender and has absorbed all the liquid. Stir with the prongs of a fork to help separate the grains; if serving with hot spiced dishes you will not need to add more salt as this dish provides a refreshingly bland, cooling contrast. Serves 6.

———— • ————

FRIAR'S BEARD CHICORY WITH LEMON DRESSING

1½ lb (700 g) friar's beard chicory (see method)

extra-virgin olive oil

1 lemon

salt

This delicious spring vegetable looks quite unlike the more common types of chicory. If you cannot buy or grow it, try purslane, (which grows wild and can be bought from Cypriot and other ethnic grocers) or marsh samphire, which grows in salt marshes or on rocks and is gathered in Brittany and Norfolk each summer and sold by an increasing number of greengrocers and fishmongers. Samphire should be steamed until only just tender. Do not overcook.

Rinse your chosen vegetable well after trimming, drain and add to a large pan of boiling salted water, stirring with a fork or long-handled spoon. If friar's beard is used, boil hard for 8–10 minutes or until tender. Drain well and sprinkle with a little oil mixed with lemon juice (or butter).

———— • ————

FRIAR'S BEARD CHICORY WITH GARLIC AND OIL

1¼ lb (600 g) friar's beard chicory (see method)

4 tbsp extra-virgin olive oil

4 cloves garlic

1 small dried chilli pepper

salt

An almost infinite variety of green and leaf vegetables can be cooked in this way. Purslane (see previous recipe) is a good substitute for the friar's beard. Heat the oil in a large saucepan, add the whole, peeled garlic cloves and fry gently until very lightly browned. Crumble in the chilli pepper, stir, then add the friar's beard chicory or substitute. Stir-fry over a high heat for about 5 minutes, until tender but still with a bit of crispness left. Sprinkle with a pinch of salt, stir once more and serve piping hot.

———— • ————

POACHED TURNIP TOPS

4½ lb (2 kg) turnip tops or purple sprouting broccoli

1 large Spanish onion

12 fl oz (350 ml) chicken stock (see page 37)

3½ fl oz (100 ml) extra-virgin olive oil

salt and peppercorns

Serve with:

slices of mild, semi-hard cheese

crusty wholemeal bread

Trim the turnip tops, removing the larger stalks and ribs; if using sprouting broccoli, cut the larger pieces in half. Wash well. Peel the onion, cut lengthwise in half and slice thinly.

Place the turnip tops or broccoli in a large, heavy-bottomed fireproof casserole dish, add the onion, stock, a pinch of salt of 3–4 black or white peppercorns. Bring to the boil, cover and simmer gently for 15 minutes (10 minutes for broccoli).

Sprinkle the olive oil all over the turnip tops, turn up

the heat and cook uncovered for up to 10 minutes (broccoli will need only about 5 minutes), or until the turnip tops are done but still have a little bite left to them. Serve with the reduced cooking liquid as a sauce, and eat with slices of semi-hard cheese and wholemeal bread.

———•———

CELERY AU GRATIN

2 heads white celery weighing approx. 2 lb (900 g) in total

7 fl oz (200 ml) chicken stock (see page 37)

2 tbsp dry vermouth or dry white white

1 shallot, peeled and finely chopped

1 sprig fresh thyme

¼ tsp cornflour or potato flour

2 oz (60 g) butter

1½ oz (40 g) grated Gruyère or Emmenthal cheese

1½ oz (40 g) freshly grated Parmesan cheese

salt and white pepper

Preheat the oven to 400°F (200°C) mark 6. Cut off the hard, woody bases and the leaves of the celery bunches, separate the stalks and run a potato peeler down the outside of each stalk to get rid of any strings. Cut the stalks into 1½-in (4-cm) lengths, place in a wide, fairly shallow saucepan or frying pan, add the stock, cover tightly and cook for about 7 minutes or until tender but still crisp.

Drain off the liquid into a small saucepan and add the shallot and thyme; boil until the liquid has reduced by half. Mix the cornflour or potato flour with a little cold water, stir into the liquid and then simmer for 1 minute while stirring the slightly thickened liquid.

Grease a gratin dish with butter and spread out the celery in a single layer.

Melt 1½ oz (40 g) butter in a small saucepan and sprinkle it all over the celery. Sprinkle with the grated cheeses and a little freshly ground pepper. Drizzle the slightly thickened stock all over and place in the oven for about 10 minutes.

CELERY JAPANESE STYLE

Selori no Kimpira

2 large, wide green celery stalks

1½ tbsp sunflower oil

1 tbsp Japanese soy sauce

2 tbsp saké or dry sherry

pinch caster sugar

pinch ajinimoto (Japanese taste powder) (optional)

Trim off the ends of the celery stalks and their leaves. Scrub well under running cold water. Use a potato peeler to remove the outer layer, making sure there are no strings left. Cut lengthwise in 4 equal strips, then cut these into large matchstick pieces about 1¾ in (5 cm) long.

Heat the oil in a non-stick frying pan and stir-fry the celery for 1½ minutes. Sprinkle with the soy sauce and saké, turn up the heat a little more and reduce what little liquid there is so that the celery is just moist.

Add a pinch of sugar and a pinch of the taste powder, if wished. Stir once more and serve on 4 individual plates. This is very good cold or hot, served with plain boiled or steamed rice.

———•———

FENNEL AU GRATIN

6 bulbs fennel

1 clove garlic

1 sprig thyme

1½ oz (40 g) grated hard cheese

approx. 3 oz (90 g) fine fresh breadcrumbs

2 oz (60 g) butter

salt and pepper

Serve with:

Rösti Potatoes with Shallots (see page 196)

Broccoli Purée (see page 93)

Preheat the oven to 400°F (200°C) mark 6. Heat water in the lower compartment of a steamer. Prepare the fennel, wash and cut lengthwise into wide, thick slices. Steam for 10 minutes or until tender but firm. If preferred, boil the fennel until tender.

Melt 1½ oz (40 g) butter in a wide, fairly shallow fireproof and ovenproof *sauteuse* or casserole dish and fry the fennel slices lightly with a clove of garlic and a sprig of thyme. Season with a little salt and freshly ground pepper and brown lightly on both sides. Sprinkle with the grated cheese and a thin covering of breadcrumbs, dot flakes of butter over the surface and place in the oven for 5 minutes.

Serve at once as a main dish or vegetable accompaniment to fish or poultry dishes.

———•———

CURRIED FENNEL

6 bulbs fennel
1 clove garlic
1 sprig fresh thyme
7 fl oz (200 ml) chicken stock (see page 37)
1½ tbsp mild curry powder
1½ oz (40 g) grated hard cheese
2 oz (60 g) fine breadcrumbs
2 oz (50 g) butter
salt and pepper

Set up the steamer ready for use, with the water heating in the lower compartment. Preheat the oven to 350°F (180°C) mark 5. Prepare the fennel, rinsing well, dry and slice lengthwise into wide, fairly thick slices. Steam for approx. 10 minutes or until just tender. Alternatively, boil until tender. Melt 1 oz (30 g) butter in a frying pan and fry the fennel gently on both sides with the peeled garlic clove and thyme sprig for about 5 minutes to brown lightly.

Add a little salt and freshly ground pepper and the stock mixed with the curry powder. Cover and simmer for 5 minutes.

Transfer to a buttered gratin dish, sprinkle with the cheese and breadcrumbs, dot a few flakes of butter here and there and brown in the oven for 10 minutes.

This dish can be prepared in advance, sprinkled with the topping shortly before placing in the oven and given a little longer to heat through and brown.

———•———

CAULIFLOWER AND POTATOES INDIAN STYLE
Aloo Gobhi

1 small cauliflower
2 cold boiled medium-sized waxy potatoes
1½ tsp ground cumin
1 tsp cumin seeds
1 small green chilli pepper, seeds removed
½ tsp ground coriander
½ tsp turmeric
5 tbsp sunflower oil
salt and pepper
Serve with:
chapatis (see page 81) or nan (Indian flat bread)

Prepare the cauliflower: cut off the larger stalks and solid root and divide the head into small florets. Cut the potatoes into small cubes. Toast ½ tsp of the ground cumin in a non-stick pan with no added oil over a low heat for 1 minute while stirring; set aside. Heat the oil in a large non-stick frying pan and cook the cumin seeds in it for about 3 seconds. Add the cauliflower florets and stir-fry over a moderately high heat. Reduce the heat to very low, cover and cook for 7 minutes by which time the florets should be just tender. Mix 1 tsp salt with the chopped chilli pepper, the toasted ground cumin, the ground coriander, turmeric and freshly ground black pepper in a small bowl.

Add the potatoes to the cauliflower, stir in the spice

mixture and cook for 4 minutes over a low heat stirring carefully. Serve at once, with chapatis, or heated nan bread brushed with melted butter.

———•———

CAULIFLOWER SALAD

1 cauliflower

6 gherkins pickled in mild vinegar

9 green olives, stoned

9 black olives, stoned

6 small anchovy fillets

1 pickled or canned sweet pepper

2 tbsp capers

extra-virgin olive oil

wine or cider vinegar

salt and pepper

Bring a large saucepan of salted water to the boil. Cut the cauliflower (see page 19) into quarters and add to the boiling water to cook for 15–20 minutes or until tender but still firm. Drain immediately and leave to cool.

Cut the gherkins into thick slices, the anchovies into small pieces and the drained sweet pepper into short, thin strips. Cut off the larger stalks and 'core' of the cauliflower and divide the head into individual florets. Place these in a large salad bowl with all the other ingredients and sprinkle with a dressing made with a pinch of salt and freshly ground pepper mixed with 1½ tbsp good wine vinegar or cider vinegar and approx. 3 fl oz (90 ml) best-quality olive oil. Stir carefully to dress without breaking up the cauliflower florets.

Do not use gherkins, sweet peppers or capers that have been pickled in acetic acid as this is so sharp and strong that it will effectively drown every other flavour.

SWEET-SOUR CHINESE LEAVES PEKING STYLE

1 head Chinese leaves weighing approx. 2 lb (1 kg)

1 small piece fresh ginger, peeled and finely chopped

1 dried red chilli pepper

4 tbsp sunflower or peanut oil

3½ fl oz (100 ml) stock

salt

For the sauce:

1½ tbsp plain flour

1½ tbsp light soy sauce

3 tbsp wine vinegar or cider vinegar

3 tbsp/¼ cup Chinese rice wine, saké or dry sherry

2 tbsp caster sugar

4 tbsp unsweetened orange juice

½ light stock cube, crumbled

Serve with:

steamed rice (see page 126)

Cut off the hard base of the Chinese leaves and peel off the outer, wilted leaves. You will need about 1¾ lb (800 g) of crisp, tender leaves. Rinse the leaves well, drain and cut across into pieces about 1½ in (4 cm) wide.

Mix the flour with the soy sauce in a small bowl; stir in all the other sauce ingredients as listed above.

Heat the oil in a large wok or non-stick frying pan and fry the chopped ginger for a few seconds before crumbling in the chilli pepper and adding the Chinese leaves. Stir-fry over a high heat for 2–3 minutes.

Add the stock, stir, cover and reduce the heat to very low. Simmer for 5 minutes, stir the sauce ingredients once more and then mix into the contents of the wok. Turn up the heat a little and stir for 1 minute or until the sauce has thickened and coated all the leaves; add a very little salt if wished. Serve at once. This is delicious with plain steamed rice.

BRAISED RED CABBAGE WITH APPLES
Rotkraut

---

1 red cabbage

4 firm dessert or cooking apples

approx. 3 tbsp red wine vinegar

1 small onion

1 oz (30 g) butter

1 tbsp oil

7 fl oz (200 ml) red wine

salt and black peppercorns

Prepare the cabbage and cut into quarters, removing and discarding the tough remains of the stem and the very large ribs. Cut lengthwise into thin strips. Peel, quarter, core and slice the apples and mix the cabbage and apples in a large mixing bowl, sprinkling them with the vinegar and a generous pinch of salt. Stir.

Peel the onion and chop finely; fry gently in the butter and oil in a large, heavy-bottomed saucepan or casserole dish (preferably stainless steel or enamelled cast iron). When the onion is tender but not browned, add the shredded cabbage together with the apple slices. Stir well and add the wine. Cover with a tight-fitting lid and simmer gently for 40 minutes, adding a little water if necessary. When the cabbage is tender, add salt if needed and plenty of freshly ground black pepper, stir once more and serve very hot.

BRUSSELS SPROUTS WITH TOASTED SESAME SEEDS

1¼ lb (600 g) Brussels sprouts

2 tbsp white sesame seeds

1 oz (30 g) butter

salt and white pepper

Bring a large saucepan of water to the boil. Add a generous pinch of salt. Prepare the Brussels sprouts and add to the boiling water to boil fast, uncovered, for up to 15 minutes or until they are just tender. Drain in a colander.

Toast the sesame seeds without any oil or fat, spread out in a large non-stick coated frying pan for 1–2 minutes over a moderate heat, stirring continuously to prevent them overcooking or burning. Remove from the heat. Heat the olive oil in the emptied saucepan used to boil the Brussels sprouts, add the sprouts and stir gently over a low heat for 2–3 minutes. Transfer to a heated serving dish, sprinkle with the sesame seeds and serve at once.

BROCCOLI PURÉE

2¾ lb (1.2 kg) sprouting broccoli or calabrese

3 shallots or 1 medium-sized mild red onion

½ clove garlic

1½ oz (40 g) butter

approx. 4 fl oz (120 ml) chicken stock (see page 37)

nutmeg

1 oz (30 g) grated Gruyère or Emmenthal cheese

5 fl oz (150 ml) double or soured cream

salt and pepper

Trim and prepare the broccoli, discarding all the larger, older stalks. Cut off the budding flower heads or small florets; chop the stalks coarsely.

Peel the onion and chop it finely together with the garlic; fry these gently in ½ oz (15 g) of the butter until soft, add the chopped small stems and cook, stirring, for 2–3 minutes. Pour in the stock, cover and simmer over a low heat for 10 minutes or until the stalks are very tender and the liquid has com-

93

pletely evaporated. Reduce to a purée in the food processor, adding a pinch each of salt, pepper and freshly grated nutmeg.

Boil the florets in salted water until just tender, turn into a colander or sieve and refresh under running cold water. Drain well before chopping finely.

Heat the purée just before you plan to serve it, stirring in the chopped florets. Draw aside from the heat when hot, stir in 1 oz (25 g) butter, the grated cheese, a little more salt and pepper if needed and the cream.

---·---

RICE WITH SAVOY CABBAGE

8 large, crisp Savoy cabbage leaves
3 medium-sized waxy potatoes
7 oz (200 g) risotto rice (e.g. arborio)
4 fairly thick rashers streaky bacon
3–4 sage leaves
1 oz (30 g) butter
1 tbsp oil
5 tbsp grated hard cheese
salt and pepper

Bring a large saucepan two-thirds full of water to the boil with a pinch of salt. Rinse the Savoy cabbage leaves and shred into a chiffonade (see page 24). Peel the potatoes and dice. Add the cabbage and potatoes to the boiling water and cook fast for 2 minutes. Sprinkle in the rice and cook for about 14 minutes or until it is tender but still has a little bite to it. Drain and leave to cool. This first stage of preparation is best completed several hours in advance; the mixture is easier to handle if chilled in the refrigerator; take it out for about 20 minutes before completing the recipe.

Dice the bacon and fry in the hot butter and oil in a large saucepan with the sage leaves for 2–3 minutes while stirring; the bacon should become crisp and brown. Do not allow the butter to burn. Add the rice and vegetable mixture, turn up the heat and fry while stirring and turning for 2–3 minutes to reheat and flavour. Remove from the heat, sprinkle with the grated cheese and serve at once on hot plates.

---·---

CARDOONS AU GRATIN

approx. 2 lb (1 kg) cardoons (see method)
1 lemon
vegetable stock (see page 38)
2 oz (60 g) small slivers of Parmesan cheese
2 oz (60 g) small slivers of Gruyére or Emmenthal cheese
2 oz (70 g) butter
2½ oz (70 g) white truffle, fresh or tinned (optional)
salt and white pepper

Have the vegetable stock prepared in advance. Reheat it slowly over a low heat, stirring now and then. Trim and wash the cardoons; you will need to remove the outer, stringy layer of the stalks with a potato peeler. Cut into 2-in (6-cm) sections and drop them straight into a bowl of iced water mixed with the juice of 1 lemon as you prepare them or they will discolour. When the cooking liquid has come to the boil, drain the cardoons and add to it. The cooking liquid should completely cover the vegetables. Drain well.

Preheat the oven to 400°F (200°C) mark 6. Grease a wide, shallow ovenproof dish with butter and place a layer of the cooked vegetables in it. Cover with half the mixed cheeses (use a mandoline cutter if you have one to slice the cheeses wafer thin). Dot half the remaining butter in flakes over the surface. Cover with a second layer of the vegetables, followed by the cheeses and remaining butter. If you are using a truffle or tinned truffle peelings, sprinkle these on top and bake in the oven for 10 minutes or until the surface has turned a pale golden brown. Serve at once.

SWEET SPINACH PIE

2–2½ lb (1 kg) spinach
1 piece fresh ginger, very finely chopped
1½ oz (35 g) butter
1 oz (30 g) sultanas
3½ fl oz (100 ml) dark rum
5 eggs
4 oz (120 g) soft brown cane sugar
4 allspice berries, ground
nutmeg
1 small lemon
1 oz (30 g) pine nuts
10 fl oz (300 ml) double cream
2 8-oz (250-g) packets frozen puff pastry, thawed
1½ tbsp sesame seeds
Serve with:
6 fl oz (180 ml) whipping cream, fairly stiffly beaten

Trim off the spinach stalks and keep them for use in other dishes; wash the leaves, do not dry or shake them, but place directly in a non-stick coated saucepan with no additional water.

Cover and cook over a moderate heat for 2 minutes, turning the spinach once. Transfer to a colander, refresh under running cold water; drain and squeeze very tightly to get rid of as much moisture as possible. Chop finely.

Peel and chop the ginger; heat 1 oz (25 g) of the butter in a non-stick coated frying pan and cook the ginger for a few seconds, stirring it constantly; add the spinach and cook for 2 minutes, stirring continuously. Set aside to cool.

Soak the sultanas in the rum for 15 minutes.

Beat 4 of the eggs in a very large bowl with the sugar; add the allspice (use a generous pinch of ground allspice if you cannot find the berries), a pinch of grated nutmeg and the finely grated rind of the lemon.

Cook the pine nuts in 1 tbsp butter in a non-stick pan for a few seconds, shaking the pan; they should be a pale golden brown. Add the cream to the egg mixture in the bowl, followed by the spinach, sultanas, rum and pine nuts. Preheat the oven to 400°F (200°C) mark 6.

Roll out the two thawed pieces of pastry into rectangles and use one to line a non-stick Swiss roll tin or rectangular baking tray. Spread all the spinach mixture evenly over the surface, leaving an uncovered border all the way round the edges. Cover with the second rectangle of pastry and seal the edges tightly, pressing them down all the way round with the prongs of a fork. Beat the remaining egg lightly and brush all over the surface of the pie. Sprinkle with the sesame seeds and bake in the oven for 20–25 minutes, when the pastry should be crisp and pale golden brown. Leave to stand for 10 minutes then serve with whipped cream.

PERSIAN RHUBARB FRAPPÉ
Sharbate Rivas

2–2½ lb (1 kg) tender pink rhubarb stalks
1 lb 2 oz caster sugar
finely chopped ice
Decorate with:
cocktail cherries
mint leaves

Wash the rhubarb stalks well and trim off both ends; if the stalks are not very thin and young, run the potato peeler down them to remove any tough fibres and cut into small pieces. Place in a heavy-bottomed enamelled fireproof casserole dish or stainless steel saucepan with 18 fl oz (500 ml) water. Bring to the boil, cover and simmer gently for 25 minutes over a low heat. Process with a vegetable mill, push through a non-metallic sieve; pour into a measuring jug and add a little water if necessary to make up a volume of 14 fl oz (400 ml). Pour back into an enamelled or stainless steel casserole dish or saucepan, add the sugar and stir over a moderate heat until it has totally dissolved. Boil the rhubarb purée gently, uncovered, for about 6 minutes or until the sugar thermometer registers 425°F (220°C) (or until a drop added to a saucer of iced water forms a malleable ball). Draw aside from the heat and leave to cool, then chill.

Just before serving, pour 4 fl oz (120 ml) of the rhubarb into large, tall glasses and fill up with chopped ice. Decorate with the cocktail cherries and mint; place the glasses on small plates with long-handled ice cream soda spoons. Alternatively, freeze the cooled rhubarb mixture in an ice cream maker or in a shallow bowl, stirring at intervals as it freezes and thickens.

———— • ————

RHUBARB CAKE

1¼ lb (600 g) rhubarb stalks (net weight)
12 oz (350 g) sugar
generous pinch cinnamon
5 oz (150 g) butter
3 eggs
8 oz (220 g) plain flour
1 tsp (1½ tsp) baking powder
4 tbsp milk
1 lemon
salt

The day before you plan to make this cake, trim and rinse the rhubarb, running a potato peeler over the surface to take off the outermost layer unless it is very young and tender. Any thick stalks should be cut lengthwise in half, so that all the pieces are of an even thickness. Cut them into 1½-in (3½-cm) sections and place in layers in a large non-metallic bowl (plastic or glass are ideal). Mix the sugar with the cinnamon and sprinkle a little on to each layer, reserving most of it for sprinkling over the top layer.

Leave to stand in a cool place overnight.

When it is time to make the cake, preheat the oven to 375°F (190°C) mark 5. Work the butter, softened at room temperature, with 5 oz (150 g) of the sugar; beat until pale and fluffy; beat in the eggs one at a time. Sift in the flour with the baking powder and a pinch of salt; stir in well, adding the milk and the juice and finely grated rind of the lemon.

Grease a 9–10-in (24-cm) flan dish or fairly shallow spring-release cake tin with butter; drain off all the liquid from the rhubarb and chop 7 oz (200 g) of the pieces finely; fold into the cake mixture and turn into the flan dish or cake tin; smooth the surface level with a palette knife. Cover with the remaining pieces of rhubarb, arranging these neatly side by side in concentric circles, working from the outer edge inwards. Bake in the oven for 45 minutes; leave to settle and cool slightly before taking it out of the cake tin (if baked in a ceramic flan dish you will not need to take it out). Serve warm or at room temperature.

·SHOOTS AND FRUITS·

Provençal stuffed vegetables

STUFFED BAKED PEPPER AND ANCHOVY ROLLS

p. 111

Preparation: 30 minutes
Cooking time: 35–40 minutes
Difficulty: very easy
Appetizer

AVOCADO, TOMATO AND GRAPEFRUIT SALAD WITH CITRONETTE DRESSING

p. 111

Preparation: 20 minutes
Difficulty: easy
Appetizer

AVOCADO AND SHRIMPS IN TUNA SAUCE

p. 111

Preparation: 15 minutes
Cooking time: 2 minutes
Difficulty: very easy
Appetizer

JAPANESE CUCUMBER AND HAM SALAD

Sunomono

p. 112

Preparation: 10 minutes + 1 hour chilling time
Difficulty: very easy
Appetizer

CUCUMBER WITH SALMON ROE AND HARD-BOILED EGGS

Ikura to Kyuri

p. 112

Preparation: 10 minutes + 20 minutes standing time
Cooking time: 10 minutes
Difficulty: easy
Appetizer

GUACAMOLE

p. 113

Preparation: 15 minutes
Difficulty: very easy
Appetizer

PILI-PILI TOMATO SALAD

p. 113

Preparation: 15 minutes + 3 hours soaking time
Difficulty: very easy
Appetizer or accompaniment

STUFFED TOMATOES AU GRATIN

p. 114

Preparation: 30 minutes
Cooking time: 15 minutes
Difficulty: very easy
Appetizer or accompaniment

COURGETTES ESCABECHE

p. 114

Preparation: 15 minutes + chilling time
Cooking time: 20 minutes
Difficulty: easy
Appetizer or accompaniment

DEEP-FRIED OKRA

p. 115

Preparation: 30 minutes
Cooking time: 20 minutes
Difficulty: easy
Appetizer or accompaniment

SPRING VEGETABLE SOUP

p. 115

Preparation: 20 minutes
Cooking time: 1 hour
Difficulty: very easy
First course

CREAM OF ASPARAGUS SOUP

p. 115

Preparation: 20 minutes
Cooking time: 40 minutes
Difficulty: easy
First cource

CASTILIAN CREAM OF ARTICHOKE SOUP

p. 116

Preparation: 20 minutes
Cooking time: 40 minutes
Difficulty: very easy
First course

CHILLED TOMATO SOUP WITH PRAWNS

Crème de Tomates Bec Fin

p. 117

Preparation: 30 minutes + 3 hours chilling time
Cooking time: 10 minutes
Difficulty: very easy
First course

CREAM OF TOMATO SOUP

p. 117

Preparation: 20 minutes
Cooking time: 30 minutes
Difficulty: very easy
First course

PUMPKIN SOUP WITH AMARETTI

p. 118

Preparation: 25 minutes
Cooking time: 30 minutes
Difficulty: very easy
First course

CHILLED PUMPKIN SOUP

Fonda del Sol

p. 118

Preparation: 20 minutes + 2 hours chilling time
Cooking time: 1 hour
Difficulty: very easy
First course

BAMBOO SHOOT AND ASPARAGUS SOUP

p. 118

Preparation: 20 minutes + 30 minutes soaking time
Cooking time: 40 minutes
Difficulty: very easy
First course

AVOCADO SOUP

Sopa de Aguacate

p. 119

Preparation: 15 minutes
Cooking time: 5 minutes
Difficulty: very easy
First course

CHINESE CUCUMBER, PORK AND CHICKEN SOUP

p. 119

Preparation: 15 minutes
Cooking time: 8 minutes
Difficulty: very easy
First course

COURGETTE SOUP WITH HERBS AND LEMON

p. 120

Preparation: 25 minutes
Cooking time: 40 minutes
Difficulty: very easy
First course

GAZPACHO

p. 120

Preparation: 30 minutes + 2–6 hours chilling time
Difficulty: very easy
First course

RISOTTO WITH COURGETTE FLOWERS

p. 120

Preparation: 15 minutes
Cooking time: 20 minutes
Difficulty: easy
First course

ARTICHOKE RISOTTO

p. 121

Preparation: 20 minutes
Cooking time: 20 minutes
Difficulty: easy
First course

SPAGHETTI WITH TOMATOES, GARLIC AND CHILLI

p. 122

Preparation: 15 minutes
Cooking time: 10 minutes
Difficulty: very easy
First course

SPAGHETTI WITH MEDITERRANEAN SAUCE

p. 122

Preparation: 15 minutes
Difficulty: very easy
First course

SPAGHETTI WITH AUBERGINES

p. 122

Preparation: 15 minutes
Cooking time: 20 minutes
Difficulty: easy
First course

SPAGHETTI WITH COURGETTE FLOWERS, CREAM AND SAFFRON

p. 123

Preparation: 15 minutes
Cooking time: 10 minutes
Difficulty: very easy
First course

PASTA CAPRI

p. 123

Preparation: 30 minutes
Cooking time: 10 minutes
Difficulty: easy
First course

TURKISH STUFFED PEPPERS

Zeytinyagli Biber Dolmasi

p. 124

Preparation: 45 minutes
Cooking time: 45 minutes
Difficulty: easy
First course

PALM HEART, PRAWN AND AVOCADO SALAD

p. 125

Preparation: 15 minutes
Cooking time: 2 minutes
Difficulty: very easy
Lunch/Supper dish

CHICKEN, TOMATO AND AVOCADO SALAD WITH TUNA SAUCE

p. 125

Preparation: 30 minutes
Cooking time: 6 minutes
Difficulty: easy
Lunch/Supper dish

Aubergine mock pizzas

Cream of tomato soup

GLOBE ARTICHOKES IN OIL

20 baby globe artichokes (or very small side buds)

juice of 1 lemon

olive oil

1¾ pints (1 litre) court-bouillon *(see page 37)*

Half fill a very large bowl with cold water and add the lemon juice to acidulate. Prepare the artichokes (see page 18), stripping them down to the heart and slicing off the leaves about three-quarters of the way down to the base. Cut lengthwise into quarters. As each artichoke is prepared, drop it into the bowl of acidulated water to prevent discoloration.

Bring the aromatic cooking liquid or *court-bouillon* slowly to the boil over a low heat and when it boils, drain the artichokes and boil gently in it until they are tender but still fairly crisp (about 12–15 minutes). Drain the artichokes upside down, then pack in layers into large, sterilized glass preserving jars; pour in sufficient olive oil to completely cover the top layer, tapping the jar on the work surface to release any trapped air bubbles. Seal tightly. Store in a cupboard at cool room temperature and wait at least 2 weeks before sampling. This method works well for other green vegetables. Serve as an accompaniment to cold meats.

———•———

SUN-DRIED TOMATOES IN OIL WITH BREAD AND CHEESE

1 lb (500 g) sun-dried tomatoes

white wine vinegar

bay leaves

olive oil

garlic

tinned anchovy fillets, drained

small, firm black olives, stoned

dried chilli peppers

oregano

capers

Serve with:

mild cheese

crusty white or wholemeal bread

Place the tomatoes in a large, non-metallic bowl, pour in sufficient vinegar to completely cover them and leave to stand for 2 hours. Drain off the vinegar and spread the tomatoes out on kitchen paper; cover with more kitchen paper and blot up excess vinegar. Use sterilized preserving jars and layer the tomatoes with all the other ingredients, starting with a small bay leaf in the bottom of the jar; cover it with a layer of tomatoes, sprinkle these with some oil, place a small, peeled clove of garlic on top, one or two pieces of anchovy fillet, a couple of olives, a piece of chilli pepper, a pinch of oregano and 2 or 3 capers. Position the small leaves against the sides of the jar as you continue layering until the jar is almost full. Make sure the olive oil completely covers the top layer and tap the jar gently to release any air bubbles.

Seal the jars tightly and store in a dark cupboard at cool room temperature. Wait at least 2 weeks before using the tomatoes. Drain shortly before serving. These tomatoes are delicious with cheese and crusty bread.

———•———

BEAN SPROUT, CUCUMBER AND CRAB SALAD
Kani to Momiyashi

12 oz (350 g) fresh bean sprouts

1 small or ½ large cucumber

7 oz (200 g) white crab meat

For the *sambai zu* dressing:

2 tbsp Japanese light soy sauce

2 tbsp Japanese rice vinegar or 1½ tbsp cider vinegar mixed with ½ tbsp water and 1 tbsp caster sugar

6 tbsp sunflower oil

salt and freshly ground white or black pepper

Bring 3½ pints (2 litres) water to the boil in a large saucepan, adding 2 tbsp salt. Rinse the bean sprouts; pick out any green seed cases and any wilted sprouts. Add to the boiling water and blanch for only 2–3 minutes. Drain quickly and leave to cool in the colander.

Rinse and dry the cucumber and cut off both ends. Cut in half in the middle of the cucumber. Use a very sharp vegetable knife to remove the skin in strips with about ¼ in (½ cm) of the underlying flesh attached to it. This will bring you close to the central, seed-bearing section which is not used for this recipe. Cut the thick 'peelings' into julienne strips (see page 24) and place them in a large bowl of iced water. Flake the crab meat coarsely and mix with the bean sprouts and well-drained cucumber strips in a large salad bowl.

Cover and chill for 1–1½ hours; transfer to individual plates and make the dressing just before serving: mix the vinegar with 2 pinches of salt and the soy sauce in a bowl. Beat in the oil, adding a little at a time so that it forms an emulsion, using a small balloon whisk or birch twig whisk. Add freshly ground pepper to taste and serve without delay.

—— • ——

ASPARAGUS WITH BÉARNAISE SAUCE

1¾ lb (800 g) large asparagus spears

For the béarnaise sauce:

2 fl oz (60 ml) white wine vinegar

2 fl oz (60 ml) dry vermouth or dry white wine

1 medium-sized shallot, finely chopped

½ tbsp dried tarragon

3 egg yolks

6 oz (180 g) butter

2 tbsp fresh French tarragon or parsley, finely chopped

3½ fl oz (100 ml) double cream

salt and black peppercorns

Clean and prepare the asparagus (see page 17) and place in the steamer, sprinkle with a small pinch of salt and steam for 10–12 minutes or until tender.

Make the sauce. Heat the vinegar, vermouth or wine, shallot, dried tarragon, a generous pinch of salt and 4 black peppercorns in a small, heavy-bottomed saucepan. Simmer, uncovered, over a moderate heat until the liquid has reduced to about 2 tbsp. Pour through a very fine sieve and set aside. Set all the remaining ingredients of the béarnaise sauce ready to use before you start cooking the asparagus: have 1 oz (30 g) of the butter well chilled and cut into pieces about the size of a hazelnut. Once the asparagus is steaming, melt the rest of the butter very gently. Beat the egg yolks until creamy in a small, heatproof bowl or in the top of a double boiler; gradually beat in the reduced vinegar mixture, adding a very little at a time; beat in half the pieces of chilled butter. Set over gently simmering water and continue beating until the mixture starts to thicken; immediately remove the double boiler from the heat and beat in the remaining chilled butter, beating in each piece before adding the next. Keep beating as you add the melted butter, starting with a few drops at a time and gradually increasing this to a very thin continuous trickle as the sauce thickens. The consistency of the finished sauce should be a little thinner than classic mayonnaise. Stir in the chopped fresh tarragon or parsley and add a little more salt and some finely ground white pepper if needed.

Beat the cream until stiff and fold into the sauce. Place the asparagus on heated plates and serve with the sauce.

STEAMED ASPARAGUS WITH MOUSSELINE SAUCE

1¾–2 lb (800 g) large, fat white or green asparagus spears

For the mousseline sauce:

14 fl oz (400 ml) hollandaise sauce (see page 85)

4 fl oz (120 ml) whipping cream

Trim and wash the asparagus (see page 17), place in the steamer, sprinkle with a very little salt and steam for about 12 minutes or until tender.

Make the hollandaise sauce; you may prefer to make this sauce after you have prepared the asparagus and before starting to cook it, so neither will be kept waiting for long. While the asparagus is cooking beat the cream until fairly stiff and fold into the hollandaise sauce.

As soon as the asparagus is done, take it carefully out of the steamer and transfer to heated plates. Spoon some of the mousseline sauce over the asparagus tips and serve at once, handing round the remaining sauce separately.

———•———

JAPANESE ASPARAGUS SALAD

1¾ lb (800 g) fresh asparagus spears

For the Japanese dressing:

3 tbsp Japanese light soy sauce

2 tsp caster sugar

2 tsp sesame oil

Clean and prepare the asparagus. For this dish only the tenderest tips 2½–3 in (7 cm) are used; the remainder can be kept for soups. Rinse and drain these tips and cut each one diagonally in half; turn one half upside down, so that the cut sides are still next to one another but with the top end of one half dovetailed against the lower end of its other half. Steam for 4–5 minutes or until tender but firm.

Drain off any moisture and arrange on small, individual plates, sprinkling each serving with a very little of the Japanese dressing. Make this by mixing the soy sauce and sugar and then beating in the oil, using a very small balloon whisk.

———•———

AUBERGINES TURKISH STYLE
Imam Bayildi

2 long, thin aubergines, each weighing approx. 14 oz (400 g)

4 medium-sized onions

4 small green peppers

½ green chilli pepper

approx. 1¼ lb (600 g) large ripe tomatoes

4 large cloves garlic, peeled

2 tsp chopped parsley

4 tbsp extra-virgin olive oil

2 tsp caster sugar

salt and pepper

Rinse the aubergines and cut off both ends. Slice lengthwise in half. Use a potato peeler to peel two strips along each half, leaving a strip of skin nearest the cut edges and along the middle. Make 3 or 4 deep lengthwise cuts along the inner, cut side of each half where the flesh is at its thickest. Sprinkle with salt and leave at an angle in a colander to allow the bitter juices to drain away from the flesh.

Slice the onions very thinly and place in a sieve resting in a bowl; sprinkle with 2 tbsp salt and mix well. Add sufficient cold water to just cover the onions, place a plate that fits easily inside the sieve on top of them, press down and place a weight on top. Leave for 30 minutes.

Blanch, peel and seed the tomatoes. Chop the flesh into very small pieces. Remove the stalks from the

107

peppers and cut them into thin rings; remove the seeds and pale inner membrane.

Rinse the aubergines thoroughly under running cold water and then dry thoroughly. Rinse the onions by pouring a kettle of near-boiling water over them, drain and squeeze out excess moisture. Mix the onions in a large bowl with the tomatoes and peppers; season with a little salt and plenty of freshly ground black pepper.

Use 1 tbsp olive oil to grease a wide, shallow flame-proof casserole dish large enough to accommodate the aubergines in one layer, cut side uppermost. Use your fingers to open out the deep incisions and push some of the onion and tomato mixture down into them; spread the rest over the exposed cut surfaces. Place a slightly crushed whole garlic clove on top of each aubergine and sprinkle the remaining oil all over the topping and garlic cloves. Sprinkle with the sugar. Pour in 9 fl oz (250 ml) cold water through a gap between the vegetables, bring to the boil, cover and reduce the heat to very low. Simmer very gently for 1¼ hours. Remove the lid and allow to cool. Sprinkle with chopped parsley.

— • —

AUBERGINE MOCK PIZZAS

2 large, round aubergines
2 medium-sized tins tomatoes
3 oz (80 g) grated Parmesan cheese
1 mozzarella cheese
oregano
extra-virgin olive oil
butter
salt and pepper

Preheat the oven to 400°F (200°C) mark 6. Oil two shallow baking sheets or trays. Cut off both ends of the aubergines, rinse and dry. Cut into round, even slices about ¼ in (7 mm) thick. Place these on the oiled trays, brush the exposed cut surfaces with olive oil and season with salt and freshly ground

pepper. Bake in the oven for 6 minutes. Chop the tomatoes and drain off excess liquid. Take the trays of aubergine slices out of the oven and place a few pieces of tomato on top, sprinkle with a small pinch each of salt, pepper and dried oregano, some Parmesan and a tiny piece of butter and cover with a slice of mozzarella cheese. The 'pizzas' can be prepared up to this point a few hours in advance if wished. Return to the oven for 15 minutes, then serve piping hot while the mozzarella cheese is still very soft.

— • —

TURKISH AUBERGINE DIP
Patlican Ezmesi

2¼ lb (1 kg) aubergines
1 small clove garlic, peeled
4 tbsp Greek yoghurt
3–4 tbsp fresh lime or lemon juice
3 tbsp extra-virgin olive oil
salt and pepper
Garnish with:
1 sprig parsley
1 black olive
Serve with:
pitta bread

Preheat the oven to 400°F (200°C) mark 6. Leave the aubergines whole. Rinse them and place on the centre shelf of the oven to bake for 30 minutes, turning them after the first 15 minutes. Take them out of the oven and leave to cool. Cut off the stalk ends, peel and cut lengthwise into quarters. If they have seeds inside, scoop these out neatly and discard. Cut the flesh into small pieces and place in the food processor with the garlic, oil, yoghurt, lime or lemon juice, salt and freshly ground pepper. Process on high speed to a very smooth purée. Transfer the dip to a serving bowl.

Decorate the top with a sprig of parsley and an olive. If prepared in advance, cover with cling film and refrigerate. Serve at room temperature, with hot pitta bread.

—— • ——

AUBERGINES JAPANESE STYLE
Yaki Nasubi

4 round medium-sized aubergines

wasabi powder (Japanese horseradish)

4 tbsp grated yamaimo (Chinese yam)

2 tbsp white sesame seeds

Japanese light soy sauce

Preheat the oven to 400°F (200°C) mark 6. Rinse the aubergines and place on the centre shelf of the oven to bake for about 30 minutes or until tender, turning them after 15 minutes. Take out of the oven and allow to cool. Slice off the ends, peel and cut lengthwise in quarters. Neatly remove any seed-bearing sections, if visible; cut the flesh into small, rectangular 'brick' shapes and heap these up on 4 small plates.

Mix 1 tbsp wasabi powder with just enough hot water to moisten and shape into 4 tiny balls; place one on each plate.

Peel a piece of Chinese yam or *yamaimo* and grate finely; you will need about 4 tbsp of its grated flesh; place a quarter of this neatly in a little heap beside the wasabi ball on each plate. Heat a non-stick frying pan and toast the sesame seeds while stirring with a wooden spoon until they start to jump. Remove from the heat immediately and crush slightly with the back of the spoon. Sprinkle over the aubergine pieces.

Hand round tiny bowls of light soy sauce. Serve this dish in the Japanese way: mix some of the wasabi with a little soy sauce on one side of the plate, then stir in a little of the grated *yamaimo*. Take hold of a piece of aubergine with chopsticks or a fork and dip in the highly flavoured relish before eating.

DEEP-FRIED GREEK AUBERGINE TURNOVERS
Melítzano Burakakia

2 large aubergines

9 oz (250 g) feta cheese

3½ fl oz (100 ml) béchamel sauce (see page 36)

5 eggs

fine dry breadcrumbs

light olive oil

salt and white peppercorns

Use a fork to crumble the feta cheese very finely in a bowl; stir in the béchamel sauce and 3 lightly beaten eggs. Season with salt and freshly ground white pepper.

Heat plenty of light olive oil in a wok or deep-fryer to 350°F (180°C) mark 4. Cut off the ends of the aubergines, rinse and dry and slice into rounds just under ¼ in (6 mm) thick. Do not use the tapering ends: the slices need to be large and flat. Deep-fry in batches for 1 minute or until very pale golden brown but not at all crisp or dry. Drain well when removing from the oil; spread out on kitchen paper to finish draining.

Beat the 2 remaining eggs lightly in a bowl with a little salt and pepper, ready for dipping the turnovers. Place about ½ tbsp of the cheese and egg mixture slightly right of centre on each slice; fold the left side of the slice over it, making the edges meet to form a pastry or turnover, pressing gently but firmly to make the surfaces adhere, enclosing the filling; use a little beaten egg to make them stick together if necessary. Egg and breadcrumb these turnovers one by one, dipping each twice in the beaten egg before coating with the breadcrumbs; as each one is coated, place ready for cooking on a very large platter or chopping board.

Deep-fry in batches in the oil for 1 minute, or until the breadcrumb coating is golden brown. Keep hot on kitchen paper to finish draining while the later batches are fried; serve at once while very hot.

GREEK AUBERGINE DIP
Melitzansoláta

2¼ lb (1 kg) aubergines

1 medium-sized onion

2 small green peppers

1 green chilli pepper (optional)

1 lemon

½ clove garlic

5 tbsp Greek yoghurt

pinch caster sugar

4 tbsp extra-virgin olive oil

salt and pepper

Garnish with:

Kalamata olives

Serve with:

pitta bread

Preheat the oven to 400°F (200°C) mark 6. Slice off the ends of the aubergines, peel them and cut lengthwise in half. Cover a baking sheet with foil and place the aubergines on it, cut sides uppermost. Bake in the middle of the oven for 30 minutes or until tender. Meanwhile, peel and chop the onion very finely. Remove the stalk, seeds and inner membrane of the green peppers and the chilli pepper if used. Chop these very finely also. Fry the onion and peppers very slowly in 2 tbsp of the oil for about 10 minutes, stirring frequently, until soft but not at all browned. Take the aubergines out of the oven when they are done and allow to cool for a few minutes before scooping out and discarding any seed-bearing sections with a spoon. Cut the aubergines into small pieces, place in the food processor with the lemon juice, the remaining oil, the peeled garlic clove, yoghurt, ½ tsp salt and a pinch of sugar; process until very smooth. Transfer this thick purée into a serving bowl, stir in the onion and pepper mixture and add a little more salt and freshly ground pepper. Chill for 30 minutes. Garnish with black olives. Serve with pitta bread.

DEEP-FRIED AUBERGINES WITH YOGHURT
Patlican Kizartmasi

1½ lb (700 g) aubergines

1 small clove garlic, peeled

10 fl oz (300 ml) natural yoghurt or Greek yoghurt

light olive oil

salt and pepper

Garnish with:

parsley

Peel the aubergines and cut into round slices about ¼ in (7 mm) thick. Spread these out on large plates, sprinkle liberally with salt and leave for 30 minutes on the drainer of your sink, having first tipped up the plates to enable the liquid they will release to drain off. Rinse the slices under running cold water and blot-dry thoroughly between pieces of kitchen paper. Heat plenty of oil to 350°F (180°C) in the deep-fryer and fry the slices a few at a time until golden brown. If preferred, shallow-fry in hot oil, turning once. Drain well, spread out on kitchen paper and sprinkle lightly with salt and freshly ground pepper. Transfer to a serving platter or individual plates.

Grate the garlic finely and mix with the yoghurt and a pinch of salt. Spoon the yoghurt over the aubergines, decorate with parsley and serve.

———•———

GREEK SALAD

4 very large ripe tomatoes

2 large and 4 medium-sized spring onions

1 cucumber

20 black olives

4 oz (125 g) feta cheese

extra-virgin olive oil

oregano

salt and pepper

Cut the tomatoes into quarters and each quarter lengthwise in half; cut these wedges across, in half, and place an equal number in 4 fairly deep individual plates. Trim the spring onions, remove the outer layer and cut the bulbs lengthwise in quarters or into eight sections if they are large. Trim the ends off the cucumber, peel and slice into thick rounds. Arrange the spring onions, cucumber and the olives in the plates and crumble a little feta cheese over each.

Sprinkle a pinch of oregano over the salads, season with salt and freshly ground pepper and drizzle some olive oil over each serving.

———— • ————

STUFFED BAKED PEPPER AND ANCHOVY ROLLS

4 large yellow or red peppers

16 tinned anchovy fillets

savoury bread stuffing (see recipe for Stuffed Tomatoes au Gratin, page 114)

10 tinned tomatoes, drained, seeds removed

extra-virgin olive oil

salt and pepper

Preheat the oven to 350°F (180°C) mark 4. Rinse the peppers and cut lengthwise into quarters; remove the stalk, seeds and white membrane. Place on a chopping board or work surface in a single layer, inside facing upwards. Drain the anchovy fillets and lay one out flat along the centre of each pepper piece; cover with 1–2 tbsp of the bread stuffing mixture, smoothing it neatly, and top with a few strips of tomato flesh. Season with salt and black pepper and sprinkle with olive oil.

Arrange the stuffed pepper quarters in a single layer in a lightly oiled shallow ovenproof dish, carefully pour in some boiling water into one corner of the dish, adding enough to barely cover the bottom and bake in the oven for 35–40 minutes. When the peppers are very tender, take out of the oven. As soon as they are no longer too hot to handle, roll them up loosely from one end to the other.

———— • ————

AVOCADO, TOMATO AND GRAPEFRUIT SALAD WITH CITRONETTE DRESSING

2 ripe avocadoes

1 ripe pink grapefruit

1 large beef tomato

3½ fl oz (100 ml) citronette dressing (see page 35)

Peel the grapefruit, cut the flesh into dice and place in a large bowl. Peel the tomatoes, remove the seeds and dice the flesh. Mix with the grapefruit.

Cut the avocadoes lengthwise in half; remove and discard the stone and place each half cut side uppermost on a plate. Fill the hollow generously with the tomato and grapefruit mixture and sprinkle liberally with the citronette dressing.

———— • ————

AVOCADO AND SHRIMPS IN TUNA SAUCE

2 ripe avocadoes

7 oz (200 g) shrimps or small prawns

1–1½ tbsp lemon juice

9 fl oz (250 ml) tuna sauce (see page 125)

extra-virgin olive oil

salt and pepper

Garnish with:

1 tbsp capers or chopped dill

Make the tuna sauce and chill in the refrigerator. Drain the shrimps and blot dry with kitchen towels. Mix with the lemon juice, 2 tbsp olive oil, a pinch of salt and freshly ground white or black pepper. Cut the avocadoes lengthwise in half, remove the stone and fill the hollow with the shrimps. Cover with a thin layer of tuna sauce. Garnish with the capers or dill and hand round the remaining tuna sauce separately.

———— • ————

JAPANESE CUCUMBER AND HAM SALAD
Sunomono

1 cucumber

2 slices lean cooked ham, just under ¼ in (½ cm) thick

scant 1 oz (20 g) fresh ginger, finely grated

4½ fl oz (130 ml) sambai zu dressing (see recipe for Bean Sprout, Cucumber and Crab Salad, page 105)

salt

Garnish with:

cucumber fans (see page 29)

Wash the cucumber and cut off the ends. Take a handful of salt and rub it hard into the skin all over; this will preserve and enhance the green colour. Wipe the skin with a clean cloth. Cut the cucumber lengthwise into quarters. Use a curved vegetable knife or tip of metal spoon to remove the seed-bearing section and a thin layer of flesh beneath it. Cut the remaining flesh and skin into regular, matchstick strips. Cut the ham into matchstick strips of the same size. Using your fingers, mix

gently together in a bowl as briefly as possible to avoid breaking up the strips.
Cover the bowl with cling film and chill in the refrigerator.
Make the sambai zu dressing and stir in the grated ginger. Cover and refrigerate for at least 1 hour.
Just before serving, transfer the salad to small bowls and decorate each portion with a cucumber fan. Serve, handing round the dressing separately.

———— • ————

CUCUMBER WITH SALMON ROE AND HARD-BOILED EGGS
Ikura to Kyuri

1 medium-sized cucumber

4 hard-boiled eggs (see method)

1 tbsp black sesame seeds

1 small tin or jar salmon roe or red lumpfish roe

½ lemon

Garnish with:

parsley sprigs

Place the eggs in a small saucepan, add enough cold water to amply cover them and bring slowly to the boil over a low heat. As soon as the water starts to boil, turn off the heat, cover the saucepan and leave for 20 minutes before draining and peeling the eggs. When completely cold, cut lengthwise in half: the yolks will not be completely hard-boiled and will be bright yellow and creamy in the centre; slice a small piece off the white curved side of the egg halves so they are stable when placed yolk uppermost on small, individual plates.
Rinse and dry the cucumbers; slice off the ends and cut lengthwise in half; cut each half in two. Scoop out the seed-bearing section neatly. Carefully cut off a thin slice all down the curved side so that they will remain stable. Cut a fairly thick layer off the top but stop about ½ in (1 cm) short of the end, so that you can bend and fold this long piece over and secure

with a cocktail stick, to represent a sail at the end to which it is still attached. If the removal of the seed section means that you cut 2 parallel strips, this will look just as attractive. Place one of these 'boats' on each plate, between the two egg halves.

Toast the sesame seeds in a non-stick frying pan over a moderate heat for 1–2 minutes, stirring with a wooden spoon; as soon as they begin to jump about in the pan, remove from the heat. Spoon them over the egg yolks.

Spread the salmon or lumpfish roe along the centre of the upper surface of the cucumber 'boats' in a thin line, squeeze a little lemon juice over them and decorate each serving with a sprig of parsley.

———— • ————

GUACAMOLE

2 ripe avocadoes
1 large ripe tomato
1 green chilli pepper
2 tbsp chopped coriander leaves
1 lime
cayenne pepper or chilli powder
1 large spring onion, finely chopped
salt and pepper

Peel the avocadoes, remove the stone and slice the flesh into a large mixing bowl. Immediately add the lime juice, or use ⅔ tbsp lemon juice, and season to taste with salt, freshly ground pepper and a pinch of cayenne or chilli powder. Mix well, crushing the avocado flesh with a fork to a smooth, thick consistency. Stir in the spring onion.

Peel the tomatoes, remove the seeds and any tough parts. Cut the flesh into small dice; remove the stalk and seeds from the fresh chilli and chop very finely. Stir the tomato, chilli and chopped coriander into the avocado. Cover and chill. Guacamole is best made not more than 1–2 hours before use. Serve as a dip.

PILI-PILI TOMATO SALAD

1¼ lb (600 g) ripe tomatoes
6 tbsp olive oil
2 tbsp finely chopped parsley
salt
For the pili-pili hot relish:
4 oz (120 g) small fresh red chilli peppers
3 cloves garlic
1 large red pepper
1 medium-sized mild onion
approx. 5 fl oz (150 ml) extra-virgin olive oil
1½ tsp salt

You can buy pili-pili relish from good delicatessens; in some versions tiny chillis are used and left whole. If you prefer to make your own relish, store it in very small batches in the freezer and thaw on the day you plan to use it as it does not keep well once thawed. Rinse the chillis, remove their stalks and the surrounding tough parts, place in the food processor with the peeled garlic cloves and process until very smooth and creamy. Rinse the red pepper and cut lengthwise into quarters; remove the stalk, seeds and white membrane. Chop coarsely. Chop the onion very finely. Add the peppers and the salt to the puréed chillis in the food processor or blender and process at high speed, adding a very little olive oil at a time in a thin trickle through the hole in the lid. Stop adding the oil once the mixture is very smooth and thick, rather like a bright red mayonnaise. Add the onion and process for 2–3 seconds only, just long enough to distribute the pieces evenly.

Transfer to very small, non-transparent freezer containers with very tightly sealing lids and freeze until required.

When you want to use the pili-pili, thaw if you have used very tiny containers, or use the tip of a strong knife to scrape off the required quantity from a still-frozen portion. Mix this with a very little olive oil and a pinch of salt and use as a very hot dip for crudités or as an added flavouring and seasoning for a wide variety of dishes or sauces; do not mix with oil and salt before adding to the latter. Pili-pili is very good in minute quantities with vegetable omelettes, and with fresh or matured cheeses.

Wash the tomatoes (use the full-flavoured, oval variety if you can buy or grow them), neatly remove the stalk and all the tough, fibrous parts (see page 14), cut lengthwise in half and discard all the seeds. Place cut side uppermost on a large serving platter and sprinkle with the parsley. Place 1–2 tsp pili-pili in a bowl with a pinch of salt and gradually beat in 6 tbsp olive oil; sprinkle all over the tomatoes.

———— • ————

STUFFED TOMATOES AU GRATIN

8 medium-sized ripe tomatoes
For the savoury bread stuffing:
1 large onion, finely chopped
3–4 oz (100–125 g) fine fresh breadcrumbs
3 oz (100 g) finely grated hard cheese
2 tbsp finely chopped parsley
1 tbsp chopped chives
olive oil
salt and pepper
Serve with:
Stuffed Baked Pepper and Anchovy Rolls (see page 111), Aubergine Mock Pizzas (see page 108), Curried Fennel (see page 91) or other stuffed, baked or braised vegetables

Preheat the oven to 400°F (200°C) mark 6. Make the bread stuffing: fry the onion very gently in 5 tbsp olive oil for 10 minutes, stirring at intervals; when tender and very pale golden brown, add the breadcrumbs and continue cooking over a low heat for a few seconds, stirring. Remove from the heat and leave to cool before stirring in the grated cheese, parsley and chives. Season to taste with salt and freshly ground pepper.

Rinse and dry the tomatoes and cut off the tops about a quarter of the way down from the stalk; scoop out all the seeds and some of the flesh. Sprinkle with a little salt and then fill the tomatoes with the bread mixture, smoothing with a knife to form a dome. Lightly oil a wide, shallow ovenproof dish and arrange the stuffed tomatoes in it in a single layer. Bake in the oven for 10 minutes. Serve hot or warm on their own as a starter, an accompaniment or as part of an assortment of other stuffed, baked vegetables.

———— • ————

COURGETTES ESCABECHE

8 courgettes
1 sprig mint
4 cloves garlic, peeled
light olive oil
red or white wine vinegar
salt

Trim off the ends of the courgettes, rinse and dry them. Slice diagonally ¼ in (½ cm) thick: spread these slices out in a single layer on a clean cloth or kitchen paper and blot dry with another cloth or kitchen paper. Heat plenty of oil in a deep-fryer to 350°F (180°C) and fry the courgette slices a few at a time until they are pale golden brown and slightly crisp on the surface. Finish draining on kitchen paper while you fry the remaining batches.

Separate the mint leaves, rinse and dry them. Chop the garlic coarsely.

Spread half the fried courgette slices out in a wide, fairly deep non-metallic dish. Sprinkle with a little salt and with half the chopped garlic. Moisten with plenty of vinegar. Place half the mint leaves, laid flat, on top of this layer. Repeat with the remaining half of the ingredients. Use enough vinegar to soak the courgettes but they should not be drenched in it. Cover with cling film and chill for several hours, preferably overnight.

DEEP-FRIED OKRA

1 lb (500 g) okra

4½ oz (130 g) chick pea flour

½ tsp ground toasted cumin seeds

½ tsp ground toasted coriander seeds

pinch chilli powder

salt

Garnish with:

coriander leaves

Mix the chick pea flour in a large mixing bowl with the spices, salt and chilli powder. Remove the stalks from the okra, rinse and dry them; cut lengthwise into thin strips. Spread out on a very large chopping board or on the work surface and sprinkle the spiced flour all over them, using your hands to lift and turn them so that they are completely coated. Heat plenty of sunflower seed oil in a deep-fryer to 350°F (180°C) and fry the okra in small batches for 2–3 minutes or until crisp and golden brown. Remove from the oil and keep warm in a low oven on kitchen paper to finish draining until the remaining batches have been fried.

Serve without delay, garnished with sprigs of coriander.

———•———

SPRING VEGETABLE SOUP

3 very small globe artichokes

5 oz (150 g) young tender spinach

5 oz (150 g) escarole

1 medium-sized onion

2–3 rashers smoked streaky bacon

3½ fl oz (100 ml) dry red wine

1¼ lb (600 g) young broad beans, shelled

10 oz (300 g) fresh peas or petits pois

4 tbsp extra-virgin olive oil

salt and black peppercorns

Serve with:

thick slices of hot French, or any crusty white, bread

Trim and wash the vegetables. Remove the outer leaves of the artichokes and cut the remaining, inner parts lengthwise into thin slices. Unless you can pick your own or buy very fresh, baby artichokes, preferably the more tender varieties, it is best to use only the bottoms; otherwise leave them out of this soup; do not use tinned ones. You could substitute tender green asparagus tips. Shred the spinach and the escarole into a chiffonade (see page 24). Chop the onion finely and cut the bacon into small dice.

Traditionally this soup would be made in a flame-proof earthenware cooking pot; use an enamelled cast iron casserole or a heavy-bottomed stainless steel saucepan. Fry the onion and bacon in 2 tbsp of the oil over a low heat, stirring, for a few minutes; when they are starting to brown, pour in the wine. Cook uncovered, until the wine has evaporated, add the beans and artichokes, stir briefly and then pour in 2 pints (1½ litres) cold water; add 1–1½ tsp salt and 5 black peppercorns.

Bring to the boil over a moderate heat, then reduce the heat, cover and simmer gently for 45 minutes. Add the peas and leaf vegetable chiffonade and simmer for 15–20 minutes more. Serves 6.

———•———

CREAM OF ASPARAGUS SOUP

2 lb (900 g) young, tender green asparagus

1¾ pints (1 litre) chicken stock (see page 37)

18 fl oz (500 ml) milk

generous 1 oz (35 g) butter

1½ oz (40 g) plain flour
2 egg yolks
5 fl oz (150 ml) single or double cream
salt and white pepper
Serve with:
croutons

Heat the milk and the stock separately. Trim and prepare the asparagus (see page 17); only the green, tender part is used for this soup. Wash very thoroughly, drain and cut into pieces about 2 in (5 cm) long.

Melt the butter over a low heat in a large, heavy-bottomed saucepan; add the flour and keep stirring as you cook it gently for 1 minute; it should not colour. Draw aside from the heat and pour in the very hot milk while beating with a balloon whisk to prevent lumps forming. Gradually stir in the hot stock, adding a little at a time.

Return the saucepan to a moderate heat and keep stirring as it heats to boiling; add the asparagus pieces. Cover and simmer over a low heat for 40 minutes. Take off the heat and beat with a hand-held electric beater until creamy (or put through a vegetable mill). Reheat for a few minutes until very hot but do not allow the soup to boil again.

Place the egg yolks in a small bowl and mix well with the cream. Take the soup off the heat and keep beating continuously as you gradually add the egg and cream thickening liaison. Add salt and pepper to taste. Serve with bread croutons fried in butter until golden brown.

— • —

CASTILIAN CREAM OF ARTICHOKE SOUP

4 very large or 8 medium-sized globe artichokes
1 pint (1.1 litres) beef stock (see page 37)
juice of 1 lemon

scant 2 oz (45 g) butter
2 oz (50 g) plain flour
9 fl oz (250 ml) milk
2½ fl oz (70 ml) double cream
2 egg yolks
salt and freshly ground white pepper
Serve with:
croutons

Heat the stock and prepare the artichokes. Prepare the artichokes, stripping off all the leaves and removing the choke. You will only be using the bottoms for this recipe. Drop each artichoke bottom into a bowl of cold water acidulated with two thirds of the lemon juice as soon as you have prepared it to prevent discoloration. Drain and slice.

Place these slices in a wide non-stick frying pan, add 5 fl oz (150 ml) of the stock, cover with a lid or foil and poach for 10 minutes.

Melt the butter in a large, heavy-bottomed saucepan; stir in the flour and cook over a low heat, stirring continuously, for 1–2 minutes. It should not colour. Draw aside from the heat and keep beating vigorously with a balloon whisk or hand-held electric beater as you gradually pour in the hot stock, adding a little at a time.

Return to a gentle heat, add the artichoke slices and their cooking juices, cover and simmer for 30 minutes. When the artichokes are very tender, beat with a hand-held electric beater or put through a vegetable mill. Stir the milk into this purée and reheat to boiling point. Season the soup with salt and pepper. Beat the egg yolks and cream together briefly in a small bowl with a pinch of salt and a little freshly ground white pepper. Gradually beat in 2 tbsp of the hot soup. Take the saucepan containing the boiling hot soup off the heat; pour the slightly thinned egg and cream liaison into the soup in a very thin stream, beating continuously.

Return the saucepan to a very low heat and keep stirring as the soup thickens slightly. It must not boil. Remove from the heat and stir in some of the remaining lemon juice.

CHILLED TOMATO SOUP WITH PRAWNS

Crème de tomates Bec Fin

2¾ lb (1.2 kg) very ripe tomatoes

1 clove garlic

1 tsp fresh marjoram leaves

pinch oregano

1 large sprig basil

3 tbsp extra-virgin olive oil

1 red chilli pepper (optional)

2–3 tbsp lemon juice

2–3 tbsp fine fresh white breadcrumbs

12–16 raw, unpeeled Mediterranean prawns

2 large shallots, finely chopped

½ oz (15 g) butter

1 tbsp light olive oil

dry vermouth or dry white wine

salt and pepper

Garnish with:

4 small sprigs basil

Use a full-flavoured variety of tomato for this soup. Blanch if necessary before peeling them, cut lengthwise in half, remove the seeds and any tough parts, and place the flesh in the blender or food processor. Process at high speed with the peeled clove of garlic, the marjoram, oregano, basil, chilli pepper (remove its stalk and seeds first), the extra-virgin olive oil, lemon juice, breadcrumbs, and a little salt and freshly ground pepper. When very smooth, taste and add more salt or pepper if needed. Process for a few seconds more. Chill for 30 minutes in the refrigerator before serving; do not chill for longer or the flavour will be blunted.

While the soup is chilling, fry the shallot very gently in the butter and 1 tbsp light olive oil for 5 minutes. Add 2 fl oz (50 ml) vermouth or dry white wine, add the prawns, cover and cook gently for 5 minutes; take them out of the pan and peel, leaving the end flippers attached. Pour the tomato soup into individual bowls and arrange 3 or 4 prawns radiating out from the centre like the spokes of a wheel. Place a small sprig of basil in the centre and serve.

———•———

CREAM OF TOMATO SOUP

1½ lb (700 g) ripe full-flavoured tomatoes (or 1 large tin tomatoes)

1¾ pints (1 litre) chicken stock (see page 37)

1 medium-sized onion, finely chopped

2 oz (60 g) unsalted butter

generous 1 oz (35 g) plain flour

juice of ½ lemon

salt and white peppercorns

Bring the stock to the boil. Meanwhile, blanch and peel the tomatoes, remove any tough parts and all the seeds. Reserve the juice for other uses. Chop the flesh coarsely and place in a bowl.

Melt 1½ oz (40 g) of the butter in a large, heavy-bottomed saucepan and fry the onion very gently for 10 minutes, stirring frequently, until soft but not browned. Add the flour and cook gently, stirring, for 1–2 minutes. Draw aside from the heat and pour in one-third of the hot stock all at once, beating vigorously with a balloon whisk until it has blended evenly with the roux mixture without forming any lumps. Pour in the remaining stock, mix well and return to a gentle heat. Add the chopped tomato and stir. Simmer for 30 minutes, covered, stirring occasionally.

Remove from the heat and beat with a hand-held electric beater until all the solid ingredients have blended into a creamy liquid. Add salt, freshly ground white pepper and lemon juice to taste. Add the remaining, solid butter in one piece and stir; serve as soon as it has melted and blended with the soup.

PUMPKIN SOUP WITH AMARETTI

1 lb (500 g) prepared pumpkin flesh

generous 1 pint (650 ml) chicken stock (see page 37)

1½ oz (40 g) butter

4 large shallots, peeled and chopped

10 fl oz (300 ml) single cream

10 Amaretti di Saronno biscuits

caster sugar

salt and white pepper

Heat the stock. Dice the pumpkin flesh. Melt 1 oz (25 g) of the butter in a large heavy-bottomed cooking pot or saucepan and sweat the shallots for 5 minutes, stirring frequently. Add the pumpkin and cook over a moderate heat for 5 minutes. Pour in the hot stock, bring to the boil then cover, reduce the heat and simmer gently for 20 minutes, or until the pumpkin is tender.

Remove from the heat and allow to cool a little; process in batches in the blender until smooth; return to the saucepan. Stir in the cream, a generous pinch each of sugar and salt and some freshly ground white pepper. Reheat to boiling point. Turn off the heat. Taste, adding a little more salt if necessary. Add the remaining butter and stir until it has melted.

Ladle the soup into heated soup bowls and sprinkle the coarsely crushed Amaretti on top. Serve at once.

———— • ————

CHILLED PUMPKIN SOUP
Fonda del Sol

1¾ lb (800 g) pumpkin (net weight)

4 ripe fresh or tinned plum tomatoes

1 medium-sized onion

2 leeks

2 vegetable or chicken stock cubes

9 fl oz (250 ml) single cream

salt and pepper

This Argentinian soup has a wonderfully rich orange colour. Peel the pumpkin, cut out all the seed-bearing sections and weigh out 1¾ lb (800 g) of flesh; cut this into small pieces. Blanch and peel the tomatoes, if using fresh; remove any tough parts and discard the seeds. Trim and peel the onion, slice lengthwise into quarters and then cut across these, into small pieces. Slice the leek.

Place all the vegetables in a large saucepan or stock pot; add scant 1½ pints (800 ml) water and the 2 stock cubes. Bring to the boil, cover and reduce the heat to low so that the liquid remains at a gentle boil. After 1 hour the pumpkin should be very tender. Remove from the heat and beat until smooth with a hand-held electric beater.

Allow to cool completely then stir in the cream; season to taste with salt and pepper, cover and chill in the refrigerator for at least 2 hours. Serve very cold.

———— • ————

BAMBOO SHOOT AND ASPARAGUS SOUP

3 oz (90 g) tinned bamboo shoots, drained

2 oz (50 g) dried bamboo shoots

1 lb (500 g) fresh, tender asparagus spears

2 pints (1.2 litres) chicken stock (see page 37)

1 lemon

1 thin piece fresh ginger, peeled

7 oz (200 g) tofu

1½ tbsp sunflower oil

2 tbsp Chinese rice wine, dry sherry or dry vermouth

1 tsp sesame oil

salt and pepper

Soak the dried bamboo shoots in plenty of very hot water for 30 minutes. While they are soaking, heat the chicken stock. Trim and prepare the asparagus, washing very thoroughly: only the tender, upper halves of the spears are used for this recipe; slice them diagonally into 1¼-in (3-cm) lengths and place in a bowl of cold water mixed with the lemon juice. Cut the tinned bamboo shoots into ¼-in (7-mm) thick slices, unless they are already sliced and then cut these into pieces approximately the same size as the pieces of asparagus. Chop the fresh ginger very finely. Drain and squeeze the soaked, dried bamboo shoots; pat-dry in kitchen paper. Cut into julienne strips (see page 24). Cut the tofu into small cubes.

Heat the sunflower oil in a large, heavy-bottomed saucepan and fry the ginger and strips of bamboo shoots briefly, until the ginger releases its aroma, then stir in the pieces of tinned bamboo shoots and fry gently for a few seconds. Add the chicken stock. Bring to the boil, reduce the heat, cover and simmer gently for 30 minutes. Add the asparagus and the tofu and simmer for a further 10 minutes, until the asparagus is tender but still a little crisp.

Draw aside from the heat, add the rice wine or substitute and the sesame oil; season to taste with salt and pepper and serve very hot. This soup is traditionally served at the end of a Chinese meal.

———•———

AVOCADO SOUP
Sopa de Aguacate

3 ripe avocadoes
7 fl oz (200 ml) single cream
1¾ pints (1 litre) chicken stock (see page 37)
2–3 tbsp dry sherry
salt and white peppercorns
Serve with:
tortillas

Bring the stock to a gentle boil over a moderate heat. While it is heating, cut the avocadoes lengthwise in half, discard the stone and peel. Cut all but one of the peeled avocado halves into pieces and liquidize at high speed with 4 fl oz of the hot stock; in a few seconds it will have turned into a smooth, thick purée. Add the cream and blend at high speed for 30 seconds more.

When the stock reaches a gentle boil, reduce the heat to very low and gradually beat in the avocado cream mixture with a balloon whisk or hand-held electric beater, adding a little at a time to blend smoothly. Do not allow to boil. Remove from the heat as soon as you have added all the avocado mixture, and stir in the sherry and salt and freshly ground white pepper to taste. Ladle into very hot soup bowls. Cut the remaining avocado half lengthwise into thin slices and place 2 or 3 of these in each bowl of soup. Serve with hot tortillas.

———•———

CHINESE CUCUMBER, PORK AND CHICKEN SOUP

1 cucumber
2 oz (60 g) very thinly sliced pork fillet
scant 1½ pints (800 ml) chicken stock (see page 37)
1 thick piece fresh ginger, peeled
2 tsp cornflour
salt and pepper

Bring the chicken stock slowly to the boil with the piece of ginger in it. Wash the cucumber and cut off the ends; run the potato peeler lightly over the skin to remove just the outermost tough layer. Cut lengthwise into quarters and then into matchstick strips. Cut the pork into very thin strips, place in a bowl and dust with the cornflour mixed with ½ tsp salt, rubbing it gently into the pieces of meat. When the chicken stock comes to the boil, add the pork and simmer gently for 4 minutes; add the cucumber and simmer for a further 3 minutes. Taste and correct the seasoning. Remove and discard the piece of ginger.

Serve very hot in heated soup bowls.

---•---

COURGETTE SOUP WITH HERBS AND LEMON

1 lb (900 g) baby courgettes

2 tbsp fresh basil leaves

2 tbsp chopped parsley

1 lemon

2 eggs

1½ oz grated hard cheese

salt and pepper

Serve with:

slices of coarse white bread crisped in the oven

Bring 2 pints (1.2 litres) water to the boil. Wash the courgettes, trim their ends off and cut them into small dice.

Heat 4 tbsp olive oil in a large, heavy-bottomed saucepan over a moderate heat; add the courgettes and brown lightly, stirring frequently. Add the boiling water and a little salt and freshly ground pepper. Bring to the boil again, reduce the heat, cover the pan and simmer for 40 minutes.

Bake the thick slices of bread in the oven until golden brown or dry fry in a non-stick frying pan. Chop the basil and parsley; grate the rind of half the lemon very finely. Beat the eggs lightly in a bowl, add the basil, parsley, cheese, grated lemon rind and a small pinch of salt.

When the soup has cooked for 40 minutes, draw aside from the heat and keep beating vigorously and continuously with a balloon whisk or hand-held electric beater as you add the egg mixture in a very thin stream; it will thicken the soup slightly.

Taste and add more salt and some freshly ground pepper before serving.

GAZPACHO

2¼ lb (1 kg) large ripe tomatoes

1 red pepper

½ mild Spanish onion

3 cloves garlic, peeled

3–4 tbsp white wine vinegar

5 tbsp extra-virgin olive oil

1 small red chilli pepper, seeds and stalk removed

3 tbsp fine fresh breadcrumbs

1 small or ½ large cucumber, chilled

5 thin slices white bread

salt

Blanch, peel and sieve the tomatoes; liquidize them in the blender at a high speed with 1 clove garlic, the chopped red pepper and onion, a pinch of salt, the vinegar, 3 tbsp of the oil, the chilli pepper and the breadcrumbs until very well blended. Add a little more salt than you would normally allow as chilling the soup masks the taste of salt. Pour into a soup tureen, cover and chill in the refrigerator for 2–6 hours. Peel and dice the cucumber.

Cut the crusts off the bread slices and cut into small cubes. Heat the remaining olive oil in a non-stick frying pan and fry the cubes over a moderate heat, turning frequently until crisp and golden brown. Sprinkle with a little salt and set aside in a bowl.

Serve the gazpacho in chilled soup bowls, handing round the bowls of cucumber and croutons separately.

---•---

RISOTTO WITH COURGETTE FLOWERS

24 freshly picked courgette or pumpkin flowers

120

14 oz (400 g) risotto rice
4 large shallots
4½ fl oz (125 ml) dry white wine
2 pints (1.2 litres) chicken stock (see page 37)
1 sprig thyme
4–6 basil leaves, chopped
2½ fl oz (70 ml) single or double cream
2 oz (60 g) grated Parmesan cheese
2 oz (60 g) unsalted butter
salt and white peppercorns

Bring the stock slowly to the boil. Peel and finely chop 3 of the shallots; rinse the courgette flowers; set 8 of them aside and slice the rest into rings.

Heat scant 1 oz (20 g) of the butter together with the oil in a large, fairly deep, heavy-bottomed frying pan. Fry the chopped shallots very gently for 10 minutes, stirring now and then; do not allow to colour. Add the sliced courgette flowers and fry over a low heat for 30 seconds, stirring gently. Add the rice, turn up the heat and cook the rice, stirring continuously, for 1–2 minutes. Add the wine and continue cooking until it has evaporated. Add 8 fl oz (225 ml) boiling hot stock, stir and continue cooking over a moderately high heat so that the liquid boils; keep adding more hot stock as the liquid in the pan reduces and is absorbed by the rice. Stir occasionally. After about 14 minutes, the rice should be tender.

While the rice is cooking, peel and chop the remaining shallot and fry with the thyme in scant ½ oz (10 g) of the remaining butter in a very small saucepan; add the remaining, whole, courgette flowers and the chopped basil leaves; cook gently, turning once or twice, for 1 minute. Stir in the cream, season with a pinch of salt and freshly grated white pepper; cook gently for another 30 seconds. The whole flowers should be still slightly crisp.

When the rice is done, draw aside from the heat, add the remaining, solid butter and the Parmesan cheese; stir gently as the butter melts. Correct the seasoning. Remove and discard the thyme sprig. In the centre of each serving place a few of the whole courgette flowers with some of the creamy sauce.

ARTICHOKE RISOTTO

4 tender globe artichokes (see method)
2 pints (1.2 litres) chicken stock (see page 37)
1 small onion
12 oz (350 g) risotto rice (e.g. arborio)
2½ fl oz (70 ml) dry vermouth or dry white wine
2 oz (60 g) unsalted butter
1 tbsp olive oil
1–2 oz (30–60 g) grated Parmesan cheese
½ clove garlic, peeled and grated (optional)
salt and pepper

Bring the stock slowly to the boil and while it is heating prepare the artichokes; you will use only the heart (the inner leaves and base); if you cannot buy very young, tender varieties of artichoke, use 12 artichoke bases or bottoms (*fonds d'artichaut*). If using artichoke hearts, cut them lengthwise in quarters, remove and discard the choke, if there is one, from the centre and slice thinly.

Peel and finely chop the onion and sweat gently in 1 oz (30 g) of the butter and 1 tbsp oil in a large, heavy-bottomed frying pan. Add the artichokes and fry gently for 5 minutes, stirring and turning them; add the rice, turn up the heat to moderately high and cook, stirring, for 1–2 minutes. Pour in the vermouth and continue stirring until it has evaporated. Pour in about 8 fl oz (225 ml) of the boiling hot stock and stir. Cook briskly, adding more hot stock when necessary and stirring at intervals. After 14 minutes' cooking the rice should be tender, with just a little bite left in it. The consistency should be very moist for this risotto; add a little more stock if necessary and cook for a minute or two longer.

Remove from the heat, add the remaining, solid butter and the Parmesan cheese; if wished, peel ½ clove garlic and grate over the risotto. Stir as the butter melts. Add a little salt and some freshly ground pepper and serve at once.

—— • ——

SPAGHETTI WITH TOMATOES, GARLIC AND CHILLI

12 oz (320 g) spaghetti

4 large ripe tomatoes

4 cloves garlic, peeled, crushed but still whole

1 dried chilli pepper (optional)

1 large sprig basil

3 tbsp extra-virgin olive oil

grated Parmesan cheese

salt and black peppercorns

Bring a large saucepan of salted water to the boil ready to cook the spaghetti. Blanch and peel the tomatoes; cut them in quarters and remove the seeds. If you cannot buy ripe, flavoursome tomatoes, use 8–10 tinned, drained Italian tomatoes. Heat the olive oil in a large, heavy-bottomed saucepan and fry the garlic cloves gently until they are pale golden brown; crumble the chilli pepper into the pan, stir and then add the tomatoes, a pinch of salt and 2 basil leaves. Cook, uncovered, over a high heat for 5 minutes, stirring a few times. Remove from the heat. Tear the rest of the basil into small pieces.

Cook the spaghetti in the boiling salted water until just tender but with plenty of bite left in them; drain, add to the tomatoes and cook for 1 minute, stirring. Turn off the heat, stir in the Parmesan cheese, plenty of freshly ground pepper and the basil.

———•———

SPAGHETTI WITH MEDITERRANEAN SAUCE

12 oz (320 g) spaghetti

For the sauce:

2 large ripe tomatoes

2 tbsp chopped parsley

2 large sprigs basil

4 large cloves garlic

generous pinch oregano

1½ tbsp capers

16 black olives, stoned and chopped

1 dried chilli pepper

5 tbsp extra-virgin olive oil

salt and pepper

Heat a large saucepan of salted water in which to cook the spaghetti.

Make the sauce. Peel the tomatoes and chop; save all their juice and transfer both juice and tomatoes to a bowl; use half a medium-sized tin of Italian tomatoes if preferred. Add the chopped parsley, the basil leaves torn into shreds, the peeled and crushed garlic cloves, the oregano, capers, the chopped olives, crumbled chilli pepper, a pinch of salt and some freshly ground pepper. Stir in the olive oil and leave to stand for a few minutes. When the water comes to the boil, add the spaghetti and cook until tender but still firm. Drain, add the sauce and stir. This sauce is also delicious with steamed brown rice.

———•———

SPAGHETTI WITH AUBERGINES

12 oz (350 g) aubergines

1 lb (500 g) ripe tomatoes or 1 large tin tomatoes

½ green chilli pepper

2 large cloves garlic, peeled

3½ fl oz (100 ml) dry white wine

pinch oregano

12 oz (350 g) spaghetti

1 oz (30 g) grated hard cheese

few basil leaves

2 tbsp extra-virgin olive oil

oil for frying

salt and black peppercorns

Garnish with:

basil leaves

Bring plenty of salted water to the boil in a large saucepan for the spaghetti. Slice the ends off the aubergines, peel them and cut lengthwise into slices then into strips. Deep- or shallow-fry briefly in batches in very hot oil until pale golden brown; take out of the oil with a slotted spoon or ladle and finish draining on kitchen paper. Season.

Blanch, peel and seed the tomatoes; cut the flesh into small pieces. Remove the stalk and seeds from the chilli pepper and chop finely; fry gently with the crushed garlic in the extra-virgin olive oil for a few seconds. Add the tomatoes and a pinch of salt and cook briskly, uncovered, for 7 minutes, stirring frequently; the sauce should thicken as the water content evaporates. Add the wine and stir while continuing to cook over a moderately high heat until the sauce has thickened again. Stir in a good pinch of oregano and remove from the heat.

Cook the spaghetti until tender but still with a little bite to them. Drain and return to the pan; add the tomato sauce and 1 oz (30 g) grated hard cheese. Season with more freshly ground black pepper. Serve without delay in hot, deep plates, with the aubergine slices spread out on top and a sprig of basil to garnish each serving.

———— • ————

SPAGHETTI WITH COURGETTE FLOWERS, CREAM AND SAFFRON

24 freshly picked courgette or pumpkin flowers

12 oz (320 g) wholemeal spaghetti

4 large shallots

1½ oz (40 g) butter

7 fl oz (200 ml) single or double cream

1 sachet or generous pinch saffron threads

salt and pepper

Heat a large pan of salted water in which to cook the spaghetti. While it is heating, prepare the flowers; cut off their stalks, insert your longest, middle finger into each flower and carefully break off the yellow, fleshy pistil and discard it. Rinse the flowers gently in cold water. Drain and spread out to dry on a clean cloth or kitchen paper. Peel and finely chop the shallots; sweat gently over a low heat in the butter in a frying pan. Do not allow to colour.

When the water reaches a fast boil, add the spaghetti and cook for about 10 minutes.

When the spaghetti have been cooking for 5 minutes cut across the flowers, slicing them into rings, add to the shallot and cook gently for 30 seconds. Add the cream and saffron and simmer, uncovered, over a slightly higher heat for 1 minute. Add salt and freshly ground white pepper to taste. As soon as the spaghetti are cooked, drain them and gently stir in the sauce. Serve at once.

———— • ————

PASTA CAPRI

12 oz (350 g) pasta shapes (e.g. quills)

2 cloves garlic, peeled

1 tbsp extra-virgin olive oil

1 dried red chilli pepper

14 oz (400 g) fresh or tinned tomatoes, chopped

1 large sprig basil

12 oz (350 g) aubergines (net prepared weight)

vegetable or sunflower oil for deep-frying

5 oz (150 g) mozzarella cheese

salt and black peppercorns

Heat plenty of salted water in a large saucepan in which to cook the pasta. Keep the garlic cloves whole but crush until almost flat by pressing a heavy knife on them; fry gently in the olive oil in a large, heavy-bottomed saucepan until pale golden brown. Add the crumbled chilli pepper (discard the seeds if you want the dish to be less peppery); stir in the tomatoes, and 2 basil leaves. Season lightly with salt and freshly ground black pepper. Bring to a gentle boil and then simmer, uncovered, for 15–20 minutes to allow the sauce to reduce and thicken considerably. Stir occasionally. While the sauce is simmering, slice the ends off the aubergines, peel and cut them lengthwise into slices about ¼ in (7 mm) thick; cut these in turn into rectangles about 1¼ × ¾ in (3 × 2 cm). Heat the frying oil in the deep-fryer to 350°F (180°C) and fry the aubergine slices in batches until pale golden brown; alternatively shallow-fry, turning once. Drain well, spread out on kitchen paper to finish draining and sprinkle with a little salt.

Cut the mozzarella cheese into small cubes or dice; tear all but a few of the remaining basil leaves into thin strips. Cook the pasta until only just tender; drain and add to the tomato mixture and stir over a moderately high heat for a few seconds. Add the aubergine and mozzarella pieces and stir over a slightly lower heat for a few seconds more, until the mozzarella melts. Remove from the heat, stir in the shredded basil leaves and serve in deep, heated plates, garnished with the reserved whole basil leaves and sprinkled with a little more freshly ground black pepper.

———•———

TURKISH STUFFED PEPPERS
Zeytinyagli Biber Dolmasi

4 medium-sized green or red peppers
1 lb (500 g) onions
1½ tbsp chopped parsley
1½ tbsp chopped fresh dill
7 oz (200 g) long-grain rice
2½ tbsp pine nuts
¾ tsp ground cinnamon
1 tbsp dried mint
1 lemon
5 tbsp olive oil
salt and pepper

Preheat the oven to 350°F (180°C) mark 4. Wash and dry the peppers; cut off a lid at the stalk end, reserving these lids. Remove all the seeds and inner membrane. Turn the peppers upside down on a chopping board.

Peel and finely chop the onions. Chop the parsley together with the dill (use 2–3 pinches of dried dill if fresh is unavailable). Rinse the rice in a sieve under running cold water until the water is completely clear. Drain.

Heat 1 tbsp olive oil in a large, heavy-bottomed frying pan and cook the pine nuts over a moderate heat until they are golden brown. Remove with a slotted spoon and set aside. Add the rest of the olive oil to that remaining in the pan and fry the onions gently in it for 10 minutes, stirring frequently, until tender and pale golden brown. Add the rice and pine nuts, cook for 1 minute, stirring, then pour in 7 fl oz (200 ml) boiling water, add 1½ tsp salt and stir again. Continue stirring for a few minutes, until the water has almost completely disappeared but the rice is still very moist. Remove from the heat and leave to cool slightly; stir in the cinnamon, the crumbled dried mint, the parsley, dill, 2 tbsp lemon juice and plenty of freshly ground black pepper. Add a little more salt if wished. Stuff the peppers with this mixture, and replace their lids.

Lightly oil a wide, shallow ovenproof dish, just large enough to accommodate all the peppers standing upright, packed close together in a single layer. Pour 9 fl oz (250 ml) boiling water into the dish between the peppers and bake uncovered in the oven for 30 minutes. Take the dish out of the oven, baste the peppers with the cooking liquid, cover with foil and return to the oven for a further 15 minutes. Serve warm or cold.

PALM HEART, PRAWN AND AVOCADO SALAD

10 oz (300 g) peeled prawns

14 oz (400 g) tinned palm hearts

2 avocadoes

12 cherry tomatoes

For the curry dressing:

2 tsp mild curry powder

½ tsp chilli powder

3 tbsp fresh lime juice

5 tbsp sunflower oil

3½ fl oz (100 ml) single cream

2 tbsp chopped coriander leaves

salt and pepper

Garnish with:

coriander leaves

If the prawns are raw, steam them for 2 minutes and leave to cool. Drain the palm hearts and slice into rings.

Cut the avocadoes lengthwise in half, remove the stone and peel them; slice lengthwise, then cut each of these slices in half. Rinse and dry the cherry tomatoes.

Arrange the vegetables in sections, the tomatoes to one side, the palm hearts in the middle and the avocado slices to the right; leave space on the far side of each plate for a small mound of prawns. Make the dressing: mix the curry powder and chilli powder with the lime juice, ½ tsp salt and some freshly ground pepper. Gradually beat in the sunflower oil, adding a little at a time so that it emulsifies evenly with the other ingredients. Beat in the cream. Stir in the chopped coriander leaves. Sprinkle this dressing over each portion of prawns, garnish with a sprig of coriander and serve at once, before the avocado has time to discolour.

CHICKEN, TOMATO AND AVOCADO SALAD WITH TUNA SAUCE

14 oz (400 g) boneless chicken breasts

1 lemon

3½ fl oz (100 ml) chicken stock (see page 37)

2 fl oz (50 ml) dry white wine

4 ripe tomatoes

2 ripe avocadoes

4 oz (120 g) tender shoots asparagus chicory

For the tuna sauce:

4½ oz (130 g) tinned tuna

1 egg

½ tsp mustard

7 fl oz (200 ml) light olive oil

2 tbsp lemon juice

dash Worcester sauce

2 tbsp white wine vinegar

3 fl oz (80 ml) dry white wine

1 tbsp capers, finely chopped

Garnish with:

chopped capers

basil leaves or parsley

Season the chicken breasts with salt and freshly ground pepper and sprinkle with a little lemon juice on both sides. Leave to marinate for 10 minutes, then arrange in a single layer in a non-stick frying pan, add the stock, the first 2 fl oz (50 ml) dry white wine and cover. Barely simmer over a very low heat for 3 minutes, turn them and simmer for 3 minutes more; test to see whether they are done by inserting the point of a sharp knife deep into the thickest part; if the juice comes out at all pink, cook for another minute on each side. Transfer the chicken to a chopping board and leave to cool.

Make the sauce: break the whole egg into the blender, add the mustard and a pinch of salt. Process at high speed, gradually drizzling in the olive oil through the hole in the blender lid. Within about 2 minutes the mayonnaise will be thick and pale. Add scant 2 tbsp lemon juice, the Worcester sauce, 2 tbsp wine vinegar, the white wine and the drained tuna. Process until a very smooth, much thinner sauce is formed. Add a little salt if necessary and plenty of freshly ground pepper to taste. Pour the sauce into a sauceboat or bowl. Stir in two-thirds of the finely chopped capers, reserving 1 tsp.

Shortly before serving the salad, cut the chicken breasts into thin, diagonal slices. Rinse and drain the asparagus chicory; cut into short lengths. Wash and dry the tomatoes, cut lengthwise in half and then slice each half. Fan out some of these slices on the side of each plate furthest away from you, slightly overlapping. Cut the avocadoes lengthwise in half, remove the stone, peel and slice the halves lengthwise; fan out these slices in turn, allowing them to overlap the tomato slices slightly and overlap each other. Finally, arrange the chicken slices and the asparagus chicory in the same way, slightly overlapping the avocado slices, on the side of the plate nearest you. Spoon plenty of tuna sauce over the chicken and asparagus chicory. Garnish with the remaining chopped capers and the basil.

———— • ————

BEAN SPROUTS WITH PORK AND STEAMED RICE
Momyashi to Gyuniku

14 oz (400 g) bean sprouts
4 thin spring onions
1 piece fresh ginger, approx. ⅜ in (1 cm) thick
5 oz (150 g) thinly sliced pork fillet
4 tbsp sunflower oil
1 clove garlic, peeled
1 dried chilli pepper
salt

For the sauce:
3 tbsp mirin, dry sherry or sweet white vermouth
2 tbsp soy sauce
4 tbsp meat stock
1 tbsp tomato purée
½ tsp caster sugar
¼ stock cube mixed with 4 fl oz (120 ml) boiling water
freshly ground pepper
½ tsp cornflour or potato flour
For the egg roll garnish:
2 eggs
1 tbsp sunflower oil
pinch caster sugar
salt
Serve with:
steamed rice (see method)

Cook the rice. Measure out the long-grain rice with a measuring cup or an ordinary cup; you will need exactly the same volume (i.e. number of cups) of cold water, plus one-quarter as much again. Place the rice in a sieve and rinse under running cold water until the water draining out of the rice runs clear. Leave the rice to soak in a bowl of cold water for 20 minutes, then drain it, transfer to a heavy-bottomed saucepan with a tight-fitting lid and add your measured quantity of cold water. Bring to the boil, reduce the heat to very low, cover tightly and simmer gently for 18 minutes without lifting the lid. Check to see if the rice is done; all the water should have been absorbed and there should be little depressions in the surface of the rice. Cover and cook for a further 2 minutes if necessary. Remove from the heat and leave to stand, with the lid on, for 10 minutes before serving. The rice will keep hot for about 30 minutes. No salt is added. Good-quality rice is needed for this absorption method to work well.

Rinse and trim the spring onions, remove the outer layer and slice the leaves and white part lengthwise in quarters and then into 1½-in (4-cm) lengths. Peel and finely chop the ginger. Cut the pork into very thin short strips. Mix all the sauce ingredients in a bowl; mix the cornflour with 1 tbsp cold water then add to the sauce ingredients. Mirin is a sweet Japanese rice wine used for cooking; you can substitute sweet white vermouth, or dry sherry mixed with a generous pinch of caster sugar. Make the egg roll garnish: beat the eggs lightly with a pinch each of salt and sugar and cook in a large, lightly oiled non-stick frying pan; the omelette must be very thin. Do not allow to brown at all. Spread out flat on a chopping board; roll up when cool and cut this roll into thin strips. Heat 2 tbsp oil in the frying pan and when very hot, stir-fry the pork strips for 1½ minutes. Stir the sauce well again before pouring all over the meat, reduce the heat and keep stirring for 30 seconds as the sauce thickens and coats all the meat. Remove the pan from the heat and set aside.

Heat 2 tbsp oil in a large wok or very large frying pan, add the peeled, crushed garlic clove, the spring onions, ginger and crumbled chilli pepper. Stir-fry over a high heat for 30 seconds; add the bean sprouts and 1 tsp salt and stir-fry for 2 minutes. Add the meat and sauce, reduce the heat a little and continue stir-frying for 1 minute more. The bean sprouts should still be quite crisp. Serve immediately with the rice, garnished with the egg roll. Serves 4 as part of an Oriental meal; 2 as a meal in itself.

———— • ————

WHITE ASPARAGUS WITH SPICY SAUCE

2¾ lb (1.2 kg) white asparagus spears
For the sauce:
4 eggs, at room temperature
1 tbsp white wine vinegar
6 tbsp light olive oil
pili-pili hot relish (see page 113), optional
salt and pepper

Place the eggs in a small, deep saucepan and add cold water; bring slowly to the boil over a low heat and when the water reaches a full boil, turn off the heat, put the lid on the saucepan and leave to stand for 30 minutes. When this time is up, drain the eggs and fill the pan with cold water; when they are completely cold, peel them.

Prepare the asparagus (see page 17), wash thoroughly and tie into 4 equal bundles, each tied with 2 pieces of string. Boil or steam for 12–14 minutes or until tender.

While the asparagus is cooking, chop the eggs finely and mix in a bowl with the vinegar, oil and a little salt and pepper. If you like a hot, peppery sauce, mix ½–1 tsp pili-pili relish with scant 1 tbsp oil and add to the sauce.

Transfer each of the well-drained asparagus bundles to a hot plate, remove the strings carefully and drizzle a little sauce over the tips; hand round the remaining sauce separately.

———— • ————

ASPARAGUS WITH EGGS AND PARMESAN

2¼ lb (1 kg) asparagus
3½ oz (100 g) butter
8 eggs
4 oz (120 g) grated Parmesan cheese
salt and pepper

This Milanese recipe makes an excellent lunch or supper dish. Prepare the asparagus as for the preceding recipe but boil or steam for only 10–12 minutes, until they are just tender and still a little crisp. Keep hot while you cook the eggs.

Place scant 1 tbsp of the butter in each of 2 non-stick frying pans and as soon as the butter has melted, before it starts to foam, break 4 eggs into each of the pans and fry gently over a low heat. Sprinkle over a little salt if wished. Heat the remaining butter in a small saucepan and as soon as it has stopped foaming, turn off the heat. Transfer the asparagus bundles to hot plates, leaving room for the eggs, remove the string and sprinkle each serving with plenty of Parmesan cheese. Drizzle the melted butter over the tips and place 2 fried eggs beside each mound of asparagus. Serve at once.

———— • ————

ASPARAGUS MOULDS

4 oz (120 g) asparagus tips, coarsely chopped
1 spring onion, chopped
2 oz (50 g) butter
1 tbsp vegetable stock (see page 38)
4 eggs
4 fl oz (125 ml) single cream
4 fl oz (125 ml) milk
1½ oz (40 g) grated Parmesan cheese
salt and pepper
nutmeg
Serve with:
Mousseline Sauce (see page 107)

Preheat the oven to 180°C (350°F) mark 4. Sweat the spring onion in 1½ oz (40 g) butter for 10 minutes. Add the asparagus tips, cook gently for 5 minutes then add the stock, cover and cook for a further 5 minutes. Reserve 2 tbsp and purée the rest in the food processor. Use the remaining butter to grease 4 6-fl-oz (175-ml) moulds or ramekins. Beat 2 whole eggs with 2 yolks, a pinch of salt, pepper and nutmeg. Beat in the cream, milk, the asparagus purée, the reserved chopped asparagus and the Parmesan. Correct the seasoning and divide the mixture between the moulds. Place these in a roasting tin and pour in enough boiling water to come half-way up the sides of the moulds. Cook for 30 minutes. Allow to stand briefly before unmoulding. Serve with Mousseline Sauce.

GLOBE ARTICHOKES ROMAN STYLE

4 very fresh young artichokes
1 clove garlic
2 tbsp finely chopped parsley
3 oz (80 g) grated mature hard cheese
best-quality olive oil
½ light stock cube
salt and black pepper

The artichokes must be young and freshly picked or they will not be tender enough for this dish. Rinse, then cut the stalks. Peel away the tough section round the base, and strip off the outer layer of the stalks. Drop into a bowl containing cold water mixed with lemon juice to prevent discoloration. Chop the stalks with the garlic and parsley. Transfer to a bowl, stir in the cheese and work in the oil a little at a time to make a thick paste, seasoning with salt and pepper.

Strip off the lower, outer leaves; cut off the top half of the artichoke; push the inner leaves apart to reach the feathery choke inside and remove this, leaving only the heart and the remaining leaves. Snip off their tips. Drain then place them upside down on a board and press down hard while turning the artichoke from side to side to make the leaves open out wide like a flower. Pack the space inside the artichokes with the prepared stuffing.

Place the artichokes the right way up in a pan large enough to contain them all snugly in one layer. Pour in sufficient water to come one-third of the way up the side of the artichokes. Sprinkle plenty of olive oil all over them. Crumble the stock cube finely and sprinkle it into the water wherever there is a gap between the artichokes. Heat until the water reaches boiling point, cover and simmer for 20 minutes. When the artichokes are tender, sprinkle with more oil and serve at once.

———— • ————

ARTICHOKE AND EGG PIE

12 tender young globe artichokes

1 lb (500 g) olive oil pastry (see recipe for Italian Easter Pie on page 50) or 8 oz (225 g) puff pastry

1 lemon

1 medium-sized onion (4½ oz/130 g), finely chopped

1 bay leaf

9 eggs

2 oz (60 g) grated hard cheese

1 small clove garlic, peeled and finely chopped

1 tbsp finely chopped fresh marjoram leaves

9 oz (250 g) ricotta cheese

olive oil

butter

salt and pepper

Make the pastry as described in the recipe for Italian Easter Pie on page 50 or use ready-made puff pastry.

Have a bowl of cold water acidulated with the lemon juice standing ready. Prepare the artichokes; if you have bought large Breton artichokes, strip off all the leaves and remove the choke, leaving only the saucer-shaped bottoms and cut these horizontally into thin round slices. Trim the more tender varieties down to the heart (see page 17), cut the hearts lengthwise in quarters then cut each quarter lengthwise into thin slices. Drop the artichokes into the acidulated water to prevent discoloration.

Sweat the onion with the bay leaf over a low heat in 1 oz (30 g) butter and 2 tbsp olive oil until tender. Drain the artichokes and blot dry with kitchen paper. Add to the onion and fry gently over a slightly higher heat for 10 minutes, turning frequently, until they have browned lightly. Season, add 2 tbsp water, cover and cook for a further 5 minutes or until the artichokes are tender. Remove from the heat and leave to cool slightly.

Beat 5 of the eggs in a large bowl with a little salt and pepper. Stir in the grated cheese, garlic, marjoram, ricotta and the artichokes.

Preheat the oven to 475°F (240°C) mark 9. Roll out the pastry and assemble the pie (see page 50). Once you have the third layer of pastry in place, spread the filling out evenly on top of it; make 4 well-spaced depressions with the back of a tablespoon, each just large enough to take a raw egg broken into it. Season these with salt and pepper. If you use puff pastry, you will only need to line the tin and cover it with a lid. Glaze the surface with beaten egg.

Complete the preparation as directed for Italian Easter Pie and bake in the oven for 30 minutes.

———•———

ARTICHOKE FRICASSÉE

12 tender young globe artichokes or 16 artichoke bottoms

vegetable stock (see page 38)

5 eggs

5 fl oz (150 ml) crème fraîche or double cream

1½ lemons

1 tbsp chopped parsley

2 tbsp chopped fresh basil

1 tbsp olive oil

1½ oz (40 g) butter

nutmeg

salt and pepper

Serve with:

thick slices of French bread, toasted

green salad

Prepare the vegetable cooking liquid. Prepare the artichokes (see preceding recipe and page 17); cook until tender but still fairly crisp (about 10 minutes, less for the artichoke bottoms) in the prepared liquid. Drain in a colander and refresh under running cold water. Drain well again.

Beat the eggs just sufficiently to blend them with the cream, 1 tbsp lemon juice, salt, freshly ground

129

pepper and a pinch of grated nutmeg. Heat the oil and butter in a wide, non-stick frying pan and fry the artichokes over a moderate heat for 2–3 minutes, turning frequently, until lightly browned; reduce the heat a little, add the beaten egg mixture and cook for only 20–30 seconds, stirring slowly until the eggs begin setting and form a thick, custard-like cream. Remove immediately from the heat, sprinkle with the parsley and basil and serve without delay on hot plates accompanied by crisp slices of toasted French bread. A crisp green salad goes well with this dish.

— • —

BAKED VEGETABLE OMELETTE

2 young tender globe artichokes or 4–6 artichoke bottoms

2 courgettes

2 oz (60 g) ceps (boletus edulis) *or cultivated closed cap mushrooms*

8 oz (225 g) spinach or Swiss chard leaves

1 lettuce

1 red pepper

1 small clove garlic, peeled

6 eggs

3½ oz (100 g) curd cheese

4 tbsp grated hard cheese

3 tbsp chopped fresh marjoram

2 tbsp light olive oil

1 oz (30 g) butter

fine dry breadcrumbs (approx. 3 oz (80 g))

salt and pepper

Preheat the oven to 230°F (150°C) mark 2. Prepare all the vegetables, cutting the artichokes into quarters, then into lengthwise slices. See page 17 and recipe for Artichoke and Egg Pie on page 129 for method of preparation. Slice the courgettes into thin rounds. Slice the mushrooms about ¼ in (½ cm)

thick. Shred the spinach or Swiss chard and the lettuce into a chiffonade (see page 24). Cut the pepper into very thin strips. Heat the butter and oil gently in a large, deep, non-stick frying pan and cook the artichokes, courgettes, spinach or Swiss chard, lettuce and garlic over a low heat for 10 minutes, stirring now and then. Add the peppers and the mushrooms and continue cooking gently for a further 5 minutes. Season with a little salt and some freshly ground pepper. Beat the eggs lightly in a large bowl with a pinch of salt and pepper; beat in the curd cheese, hard cheese and the marjoram. Lightly oil a large, round pie dish, or grease with butter; sprinkle with breadcrumbs, tipping out any excess. Stir the tender, crisp vegetables into the egg and cheese mixture and transfer to the pie dish. Sprinkle the surface with more breadcrumbs and bake in the oven for 20 minutes, or until the mixture has set.

— • —

DEEP-FRIED AUBERGINE AND CHEESE SANDWICHES

2 large aubergines

2 mozzarella cheeses

8 large basil leaves

3 eggs

fine dry breadcrumbs (approx. 4 oz (120 g))

sunflower oil for deep-frying

salt and pepper

Serve with:

tomato salad

Preheat the oven to 400°F (200°C) mark 6. Lightly oil 2 baking sheets. Slice off both ends of the aubergines where they begin to taper. Rinse, dry and slice into rounds of even thickness. Lay these out flat in a single layer on the oiled baking sheets and brush lightly with olive oil; season lightly with salt and pepper. Bake for 6 minutes, then take out and leave to cool.

When you are ready to start cooking the 'sandwiches', slice the mozzarella cheeses into ¼-in (6–7-mm) thick rounds. Beat 3 eggs lightly in a small, deep bowl with a pinch of salt and pepper. Spread the breadcrumbs out on a plate.

Place a slice of cheese on top of half the aubergine slices, press a basil leaf on top and cover with the remaining aubergine slices. Press together gently and then dip in the beaten egg to coat all over; cover with the breadcrumbs, pressing lightly.

When all the sandwiches have been dipped in egg and coated in breadcrumbs, pour oil into a non-stick frying pan; it should be about ⅝ in (1½ cm) deep. When it is hot, but not smoking hot, add the sandwiches and fry over a moderately high heat, turning once, until they are golden brown on both sides. Take out of the oil with a slotted spoon and drain briefly on kitchen paper on a very hot plate. Serve.

———•———

STUFFED AUBERGINES

2 large long aubergines (total weight approx. 1¼ lb (600 g))
1 clove garlic
1 medium-sized ripe tomato
9 firm black olives, stoned
2 tbsp capers
2 tbsp finely chopped parsley
1½ oz (40 g) grated hard cheese
wine vinegar
fine dry breadcrumbs
olive oil
salt and pepper
Serve with:
grilled chicken
green salad

Preheat the oven to 400°F (200°C) mark 6. Bring a large saucepan, three-quarters full of salted water, acidulated with 1 tbsp wine vinegar to the boil. Meanwhile, wash and dry the aubergines, slice off their ends and cut lengthwise in half. Scoop out most of the flesh, leaving behind an even layer next to the skin about ¼ in (6–7 mm) thick. Blanch these in the boiling water for 1 minute; take out with a slotted ladle and leave to drain upside down on a clean cloth. Chop the flesh together with the garlic.

Heat 3 tbsp (¼ cup) olive oil in a large, non-stick frying pan and gently fry the chopped mixture, stirring, for 2–3 minutes. Stir in 3 tbsp breadcrumbs and season with salt and pepper. Continue cooking over a low heat for a further minute, then remove from the heat.

Blanch, peel, seed and chop the tomato (use a drained tinned tomato if wished). Finely chop the olives. Stir the tomato, olives, grated hard cheese, capers and parsley into the fried mixture, adding 1 tbsp more oil. Add a little more salt if wished, and plenty of freshly ground black pepper.

Lightly oil a wide, shallow, ovenproof dish or small roasting tin; fill the aubergine skins with the mixture, sprinkle with more breadcrumbs and place carefully in the dish in a single layer. Bake in the oven for about 15 minutes and brown under a hot grill for the last 3 minutes. Serve warm or cold.

———•———

AUBERGINE, TOMATO AND CHEESE BAKE

2¼ lb (1 kg) aubergines
1¾ lb (800 g) tinned tomatoes
4 cloves garlic, peeled
1 dried chilli pepper
2 mozzarella cheeses
5 oz (150 g) freshly grated hard cheese
1 large sprig basil

131

generous pinch oregano
light olive oil
salt and pepper

Make the tomato sauce: sauté the garlic in 2 tbsp olive oil until pale golden brown; crumble in the chilli pepper (discard the seeds if you prefer a less peppery dish). Seed the tomatoes and chop coarsely; add to the garlic and chilli with a pinch of salt. Stir. Cover the saucepan and simmer gently for 20 minutes, stirring every now and then. Towards the end of this time, add a generous pinch of oregano. While the tomatoes are simmering, prepare and fry the aubergines. Trim off the ends, peel and slice lengthwise just under ¼ in (approx. ½ cm) thick. Heat plenty of oil in a deep-fryer to 350°F (180°C) and fry a few slices at a time; alternatively, shallow-fry them. They should be pale golden brown but not crisp or dry. Drain well, spread out on kitchen paper to finish draining and sprinkle with a little salt. Thinly slice the mozzarella.

Preheat the oven to 400°F (200°C) mark 6. Lightly oil a wide, shallow ovenproof dish or small roasting tin, cover the bottom with a layer of aubergine slices, sprinkle with a little of the sauce and with some of the grated hard cheese. Top with a sprinkling of some of the basil leaves, torn into small pieces, and with some mozzarella slices. Keep layering in this way until you have used all the ingredients, finishing with a mozzarella layer.

Place in the oven and cook for 15–20 minutes; finish by browning the top lightly under a very hot grill. Serve at once. Serves 6.

———•———

NEAPOLITAN STUFFED AUBERGINES

2 long large aubergines (total weight approx. 1½ lb (600 g))
2 cloves garlic

9 black olives
1–2 tbsp capers
½ tbsp finely chopped parsley
1½ tbsp finely chopped basil
3 drained tinned tomatoes, seeds removed
2 tbsp wine vinegar
approx. 3 oz (80 g) fine dry breadcrumbs
olive oil
salt and pepper
Serve with:
grilled meat, poultry or fish steaks

Preheat the oven to 400°F (200°C) mark 6. Bring a large saucepan three-quarters full of salted water, acidulated with the wine vinegar, to the boil. Rinse and dry the aubergines, cut off the ends and slice lengthwise in half. Scoop out the flesh, leaving a layer about ¼ in (7 mm) next to the skin. Blanch the hollowed skins in the saucepan of boiling water for 1 minute, drain and place upside down on a clean cloth.

Cut the flesh into small, thick slices or dice. Heat 3 tbsp olive oil in a wide, non-stick frying pan over a moderately high heat and fry 2 crushed garlic cloves together with the dried aubergine flesh for 2 minutes, stirring with a spatula. Season with salt and freshly ground pepper, cover and reduce the heat to low; cook gently for 15 minutes, adding 2 tbsp of hot water if the mixture shows signs of becoming too dry.

During this time, stone and chop the olives. Remove the aubergine mixture from the heat and stir in the olives, capers, parsley and basil, adding a little more salt and pepper if wished. Fill the drained skins with this stuffing and cover with strips of tomato flesh. Sprinkle with a very little salt and a little olive oil and top with a light covering of breadcrumbs.

Place the aubergines in an oiled ovenproof dish or small roasting tin and bake for 15 minutes. Brown under a very hot grill for 2–3 minutes. Serve hot or warm.

———•———

STUFFED AUBERGINES GREEK STYLE
Papuzakia

2 large long aubergines, each weighing about 14 oz (400 g)

1 medium-sized onion, finely chopped

8 oz (230 g) minced lean lamb

pinch chilli powder (optional)

1 small garlic clove, finely chopped

2 tbsp finely chopped parsley

9 fl oz (250 ml) tomato sauce (see pages 74–75)

4 tbsp grated mature hard cheese

olive oil

wine vinegar

salt and pepper

For the sauce:

14 fl oz (400 ml) béchamel sauce (see page 36)

2 egg yolks

2 tbsp grated mature hard cheese

Preheat the oven to 350°F (180°C) mark 4. Heat plenty of salted water in a large saucepan. Trim off the ends of the aubergines, wash and dry them then cut lengthwise in half. Scoop out the flesh, leaving a layer about ¼ in (6–7 mm) thick next to the skin.

Add 1½ tbsp wine vinegar to the boiling salted water and blanch the hollowed aubergines for 5 minutes. Drain and arrange upside down on a clean cloth or tilted chopping board to continue draining.

Heat 3 tbsp (¼ cup) olive oil in a large non-stick frying pan and fry the onion gently for about 10 minutes, stirring frequently. When it is very tender and has started to brown, add the meat and fry for 1 minute, using a wooden spoon to prevent it forming balls or lumps as it cooks. Add the chopped flesh scooped out of the aubergines. Cook over a low heat, stirring, for 2–3 minutes only. Season with salt and freshly ground pepper, adding a pinch of chilli powder if wished.

Mix the garlic and parsley and stir into the meat mixture; add the tomato sauce. Keep stirring over a low heat for 2 minutes to blend and thicken the mixture. Draw aside from the heat and stir in the grated cheese. Taste, and correct the seasoning if required.

Lightly oil a roasting tin or shallow ovenproof dish. Fill the aubergine skins with the meat and vegetable mixture, smoothing the surface. Bake in the oven for 30–35 minutes. Make the sauce: when the stuffed aubergines have finished cooking and it is time to serve them, stir the egg yolks and remaining grated cheese into the sauce off the heat. Return to a very low heat to cook, stirring gently, for a few seconds; do not allow to boil. Pour over the stuffed aubergines and serve at once.

—— • ——

AUBERGINE CUTLETS WITH TOMATO RELISH

1 very large firm aubergine

2 eggs

3–4 oz (80–120 g) fine, dry breadcrumbs

light olive oil

salt and pepper

For the relish:

2 ripe tomatoes

pinch oregano

4 basil leaves

1 clove garlic, peeled and crushed

1 tbsp extra-virgin olive oil

salt and pepper

Garnish with:

basil leaves

First make the relish. Blanch the tomatoes for 10 minutes and peel (or use drained, tinned tomatoes if wished). Remove the seeds and chop the flesh coarsely. Mix in a bowl with a little salt, freshly ground pepper, the oregano, the 4 basil leaves torn into shreds, the garlic and the extra-virgin olive oil. Peel the aubergines and cut lengthwise into thick slices. Beat the eggs lightly with salt and pepper, dip the aubergine slices and then coat all over with breadcrumbs.

Heat about 8 fl oz (225 ml) light olive oil in a very large frying pan and when very hot, fry the aubergine slices for 1 minute on each side or until the coating is golden brown; remove from the oil, draining well, and place in a single layer on kitchen paper to finish draining. Do not cover. Transfer to a large dish and serve with the fresh tomato relish mixture.

———— • ————

AUBERGINE TIMBALES WITH FRESH MINT SAUCE

1 firm aubergine (10–12 oz (300 g) in weight)
2 cloves garlic, peeled
1 sprig thyme
butter
4 eggs
5 fl oz (150 ml) single cream
5 fl oz (150 ml) milk
1½ oz (40–g) grated hard cheese
3 tbsp olive oil
salt and pepper
For the mint sauce:
1 oz (30 g) mint leaves
5 fl oz (150 ml) white wine vinegar
1 tbsp caster sugar
salt and pepper
Serve with:
Macedoine of Vegetables with Thyme (see page 196)

These quantities yield enough to fill four 7-fl oz (200-ml) or six 4½-fl oz (140-ml) dariole moulds or ramekins. Preheat the oven to 350°F (180°C) mark 4. Trim, wash, dry and peel the aubergine and dice

the flesh. Pour 3 tbsp olive oil into a large non-stick frying pan, add the garlic cloves, still whole but slightly crushed with the flat of a knife blade, and the sprig of thyme and heat. When hot, add the diced aubergine and fry in the flavoured oil over a moderate heat, stirring, for about 3 minutes. Season with salt and pepper, add 8 fl oz (225 ml) water, cover and reduce the heat. Simmer over a low heat for 20 minutes, stirring occasionally, until the aubergine is very tender. Cook, uncovered, for a further 10 minutes, stirring, to allow excess moisture to evaporate completely. Remove and discard the garlic and the thyme sprig; crush the aubergine flesh to a pulp with a potato masher to turn it into a very thick purée, firm enough to hold its shape easily. Cook for another 2 minutes over a low heat, stirring. Grease the moulds or ramekins generously with butter and have some boiling water ready in the kettle to pour into a roasting tin for a *bain-marie*.

Beat 2 egg yolks with 2 whole eggs in a bowl with a pinch of salt and pepper; beat in the cream, milk, the aubergine and the finely grated cheese. Taste and add more seasoning if necessary. Fill the moulds with this mixture, tapping their bases on the work surface to make the mixture settle evenly. Place the moulds in the roasting tin, add sufficient boiling water to come three-quarters of the way up their sides and cook in the oven for 30 minutes.

Meanwhile, make the mint sauce. Rinse the mint leaves and pat-dry on kitchen paper. Chop very finely and place in a bowl. Add the sugar, vinegar, a pinch of salt and pepper and 2 tbsp cold water.

Let the moulds stand for a few minutes after removing them from the oven before unmoulding them on to individual plates. Spoon a little mint sauce to one side of the plate and garnish each timbale with a very small mint sprig.

———— • ————

AUBERGINE PIE

2¼–2½ lb (1 kg) aubergines
5 oz (150 g) ricotta cheese

5 oz (150 g) smoked semi-hard cheese, coarsely grated
5 tbsp grated Parmesan cheese
6 tbsp fine fresh or dry breadcrumbs
1 small clove garlic, peeled and finely grated
generous pinch oregano
3–4 basil leaves or small pinch dried basil
4 eggs
nutmeg
olive oil
salt and pepper

Preheat the oven to 400°F (200°C) mark 6. Rinse the aubergines, place directly on the centre shelf of the oven and bake for 30 minutes, turning halfway through this time. Take them out of the oven and reduce the temperature to 350°F (180°C) mark 4. Trim off the ends of the aubergines and peel; cut lengthwise into quarters and remove the seeds if there are any. Cut the flesh into small pieces and transfer to a large bowl. Reduce to a smooth pulp with a fork or potato masher; blend in the ricotta, the grated smoked cheese, the Parmesan, 3 tbsp of the breadcrumbs, garlic, oregano and the basil leaves torn into small pieces.

Beat the eggs briefly with 1 tsp salt, plenty of freshly ground pepper and a pinch of nutmeg. Mix well with the aubergine and cheese mixture.

Lightly oil a 9½–10-in (24-cm) quiche dish about 1½ in (4 cm) deep. Spread the mixture out in it evenly and sprinkle the surface with the remaining breadcrumbs. Bake for 35–40 minutes.

———— • ————

CAPONATA

2 medium-sized aubergines (total weight approx. 1½ lb (700 g))
1 large green celery stalk
1 large onion, finely chopped

6 large black olives, stoned
6 green olives, stoned
2–2½ tbsp capers
2 large ripe tomatoes
3½ fl oz (100 ml) olive oil
pinch oregano
2 tbsp wine vinegar
1 tbsp caster sugar
2 cloves garlic
1 sprig fresh basil
salt and pepper

Trim and peel the aubergines, cut the flesh into small chunks and place them in a large bowl. Sprinkle with plenty of salt, stir well and leave to stand for 1 hour. Make the sauce. Trim and wash the celery and cut it into small pieces. Fry the onion gently in 2 tbsp of the olive oil for 10 minutes, stirring frequently, add the celery, olives and capers and continue cooking over a low heat, stirring, for 2 minutes. Add the coarsely chopped tomatoes, a pinch of salt, some freshly ground pepper, a pinch of oregano, the vinegar and the sugar. Cover and cook over a low heat for 20 minutes.

Drain off all the liquid that the salt will have drawn out of the aubergines, rinse them briefly and dry them thoroughly with kitchen paper. Heat a non-stick frying pan over a moderately high heat for a few minutes; when it is very hot, add the aubergine cubes and dry-fry while stirring until they have formed a dry 'skin' which will prevent them from soaking up as much oil as usual; take them out of the frying pan and pour in 3 tbsp olive oil. Crush the garlic cloves and add them to the frying pan together with the aubergine. Fry, stirring, for 3 minutes. Season with salt and pepper, add 3½ fl oz (100 ml) water, cover and cook for 8 minutes or until the cubes are tender but still firm.

Mix with the sauce and simmer over a low heat for a minute or two. Sprinkle with the basil leaves torn into small pieces and serve. Caponata may also be served chilled.

PROVENÇAL STUFFED VEGETABLES
Petits Farcis

3 small round or oval aubergines

3 young courgettes

1 small red pepper

1 small yellow pepper

6 small ripe tomatoes

3 small onions

6 large courgette or pumpkin flowers (optional)

5 oz (150 g) lean veal, thinly sliced

4 oz (100 g) trimmed chicken livers or button mushrooms

5 oz (150 g) cooked ham

1 sprig thyme

1 medium-sized garlic clove

2 tsp chopped fresh marjoram leaves

2–3 tbsp velouté sauce (see page 35)

1 egg

3½ oz (100 g) finely grated hard cheese

generous pinch oregano

olive oil

1 oz (30 g) butter

3–4 oz (80–120 g) fine dry breadcrumbs

2 tbsp wine vinegar

salt and pepper

Serve with:

French bread

Bring a large saucepan of salted water to the boil; add the vinegar to it.

Prepare and rinse all the vegetables. Cut the aubergines and courgettes lengthwise in half; pare off a thin piece from the most rounded part of the aubergines so they will be stable. Cut the peppers lengthwise in quarters, removing all the seeds, the stalk and the inner pith as usual. Cut 'lids' off the tomatoes; slice the peeled onions horizontally in half.

Rinse the freshly picked courgette or pumpkin flow- ers, remove the pistils and drain on kitchen paper or a clean cloth. If courgette or pumpkin flowers are unobtainable, use an extra pepper. Preheat the oven to 325°F (170°C) mark 3. Sauté the veal slices and the sliced button mushrooms in a little hot butter with the sprig of thyme for 2–3 minutes, turning once. Transfer to a chopping board and chop fairly finely, together with the ham.

Scoop out the flesh from the aubergines and cour- gettes, leaving a fairly thick layer next to the skin. Reserve the flesh. Carefully remove the inner part of the onion, using a grapefruit spoon or curved, small vegetable knife. Blanch these hollowed vegetables in the acidulated water, allowing 10 minutes for the onions, 5 minutes for the peppers and the cour- gettes and 3 minutes for the aubergines. Drain all these blanched vegetables upside down. Remove the seeds from the tomatoes and discard them; scoop out the flesh from the tomatoes and reserve; sprinkle the inside with a little salt.

Chop the reserved flesh from the vegetables, heat 2 tbsp olive oil in a non-stick frying pan and sauté the chopped flesh for about 5 minutes. Season with salt and pepper, cover the pan and simmer gently for 10 minutes.

Chop the marjoram leaves with the garlic and com- bine in a large mixing bowl with the chopped veal (and livers or mushrooms, if used), the velouté sauce, the lightly beaten egg, grated cheese, cooked vegetable pulp, a pinch of oregano and a little salt and pepper. If the mixture is too stiff, add a little more velouté sauce.

Lightly oil 2 baking sheets. Fill all the vegetables with the prepared mixture; if you use pumpkin flow- ers, use only 1 tsp for each and insert carefully as they tear very easily; pinch the tips of the flowers gently to enclose the filling.

Transfer the stuffed vegetables to the baking sheets, sprinkle with a topping of breadcrumbs and place a sliver of butter on each. Bake for about 30 minutes, placing each sheet in turn under a very hot grill for the final 2–3 minutes to brown.

Leave to stand for 5 minutes before serving; they are at their best warm rather than piping hot. Serve with crusty French bread warmed in the oven. Serves 6.

RATATOUILLE

2 medium-sized aubergines (total weight approx. 1½ lb (700 g))

2 large ripe or tinned tomatoes

4 courgettes

1 large onion, finely chopped

4 green peppers, finely chopped

1 fresh chilli pepper, finely chopped (optional)

7 oz (200 g) small new potatoes, scrubbed or peeled

3 cloves garlic, peeled

oregano

2 tbsp finely chopped parsley

3½ fl oz (100 ml) olive oil

salt and pepper

Serve with:

a selection of French goat's cheeses

Peel the aubergines, cut the flesh into small cubes and place these in a large bowl; sprinkle them with salt, stir well and leave to stand for 1 hour. Peel the tomatoes, remove their seeds and chop the flesh coarsely.

Trim, wash and dry the courgettes, cut them lengthwise in quarters and then cut these long pieces into 1¼-in (3-cm) lengths. Heat 4 tbsp olive oil in a very large, fairly deep non-stick frying pan and fry the courgettes with 1 crushed garlic clove for 2–3 minutes, while stirring. Sprinkle with a pinch of salt, cover, reduce the heat and simmer gently for 10 minutes or until the courgettes are tender but still a little crisp. Set aside.

Fry the onion very gently with the chilli pepper in 4 tbsp olive oil for 5 minutes in a very large, heavy-bottomed flameproof casserole dish, stirring frequently. Add the small new potatoes, turn up the heat and fry for 5 minutes, stirring and turning frequently; add the tomato flesh, 9 fl oz (250 ml) hot water, a pinch of oregano, a little salt and some freshly ground pepper. Cover, reduce the heat and simmer for 15–20 minutes by which time the pota-

toes should be tender. Stir from time to time and add a little more hot water at intervals when needed.

While these vegetables are simmering, drain and squeeze the aubergines to eliminate the liquid the salt has drawn out; dry thoroughly on kitchen paper or in a clean cloth. Heat a large non-stick frying pan without any oil in it and when it is very hot, add the aubergines and dry-fry them for 2–3 minutes over a high heat, stirring and turning the cubes continuously. This will seal them and prevent them from soaking up too much oil. Transfer them to a bowl, heat 3 tbsp oil in the frying pan and return the aubergines to it with 1 crushed garlic clove. Fry for 3 minutes, stirring and turning. Season with a very little salt and some freshly ground pepper, add 2 tbsp water, reduce the heat, cover and simmer for 5 minutes or until the aubergines are tender but not at all mushy, then transfer to the casserole dish together with the courgettes and stir over a low heat for about 2 minutes; taste and add more salt or freshly ground pepper if needed.

Remove from the heat, sprinkle with the parsley and the last garlic clove, finely chopped; the flavours are at their best when ratatouille is served warm, not piping hot. Small, fresh goat's cheeses, served cold or grilled, each sprinkled with a little olive oil and black pepper make a perfect foil.

———•———

PEPERONATA

2¾ lb (1.2 kg) large red, yellow and green peppers

3 large onions

1 fresh green chilli pepper (optional)

14 oz (400 g) ripe fresh tomatoes (or drained tinned)

2 oz (60 g) green olives, stoned

3 tbsp wine vinegar

4 tbsp olive oil

salt and pepper

Serve with:

grilled meat

crusty white, or wholemeal, bread

137

Rinse and dry the peppers, cut them lengthwise into quarters, remove and discard the stalk, seeds and white membrane, slice each quarter lengthwise into 3 broad strips. Peel the onions, slice off the ends and cut lengthwise into quarters; slice each quarter lengthwise very thinly. Discard the stalk and seeds from the chilli pepper and chop very finely. Rinse the tomatoes, cut in quarters, remove any tough parts and all the seeds.

Heat 4 tbsp olive oil in a wide, heavy-bottomed flameproof casserole dish and gently fry the onions and chilli pepper together over a low heat, stirring frequently, for 15 minutes. Add the peppers and the tomatoes and stir. Cook over a slightly higher heat for 5 minutes, still stirring frequently; season with salt and pepper and sprinkle with 3 tbsp vinegar. Reduce the heat to moderately low and cook for another 10 minutes, stirring now and then.

Add the olives and continue simmering until the peppers are tender but still fairly firm and what liquid there is has thickened. Taste, add a little more salt if necessary; turn off the heat and leave to stand for 10 minutes before serving.

———— • ————

STUFFED PEPPERS

4 large red or yellow peppers
3 oz (80 g) sliced 2–3-day-old firm white bread
10 firm fleshy black olives
3 anchovy fillets
4 basil leaves
1–1½ tbsp capers
1 clove garlic, peeled and finely grated
oregano
fine dry breadcrumbs
4 tbsp light olive oil
1 tbsp extra-virgin olive oil
salt and pepper

Preheat the oven to 400°F (200°C) mark 6. Line a baking tray or a wide shallow ovenproof dish with foil. Rinse and dry the peppers and place on their sides on the foil. Bake in the oven for about 30 minutes, turning them halfway through, until their skin has browned and lifted, making them very easy to peel. Remove from the oven and leave to cool completely. Turn down the oven temperature to 325°F (170°C) mark 3.

Peel the peppers, then cut neatly round the stalk, making an opening large enough to enable you to remove the seeds and the pale sections attached to them from the inside with two fingers, holding each pepper upside down in your other hand. Make the filling. Slice off the crusts from the bread, cut it into dice and fry these in 3 tbsp hot olive oil in a non-stick frying pan, stirring frequently, until they are crisp and golden brown all over. Drain on kitchen paper, sprinkle with a pinch of salt and place in a large mixing bowl.

Cut the stoned olives into small pieces. Chop the anchovies. Tear the basil leaves into small pieces (use a small pinch of dried basil if fresh basil is unavailable), add all these ingredients to the fried bread dice and mix; use a teaspoon to fill the peppers loosely with this mixture and then place them upright in a lightly oiled shallow ovenproof dish. Sprinkle the top of each pepper with a small pinch of salt, some fine, dry breadcrumbs and a little extra-virgin olive oil.

Bake in the oven for 35 minutes and then place under a very hot grill to brown and crisp the breadcrumb topping. Sprinkle over any juices left in the cooking dish and serve hot.

———— • ————

MEXICAN PEPPERS WITH POMEGRANATE SAUCE
Chiles en nogada

4 large green peppers
10 oz (300 g) lean minced beef

1 medium-sized onion, finely chopped

½ clove garlic, finely chopped

½–1 fresh green chilli pepper, finely chopped

2 ripe medium-sized tomatoes, peeled and seeded

1 tbsp white wine vinegar

1½ oz (40 g) sultanas

1 tsp soft light brown sugar or caster sugar

pinch ground cloves

½ tsp ground cinnamon

1 oz (25 g) slivered or flaked almonds

scant 1 oz (20 g) butter

sunflower oil

salt and pepper

For the pomegranate sauce:

4 walnut halves

1 oz (30 g) whole almonds, blanched and peeled

7 fl oz (200 ml) single cream

1½ tbsp chopped coriander leaves

pinch caster sugar

½ tsp ground cinnamon

fleshy seeds from half a pomegranate

few whole coriander leaves

1 lime

salt

Preheat the oven to 400°F (200°C) mark 6. Rinse and dry the peppers and bake in the oven for 20–25 minutes, turning halfway through to loosen their skins (see previous recipe). Peel the whole peppers carefully, remove the stalk neatly, cutting away just sufficient flesh surrounding it to make it easy to remove the seeds and pale pith using two fingers.

Make the filling while the peppers are in the oven. Heat the butter with 1 tbsp oil in a wide non-stick frying pan over a low heat and sweat the onion, garlic and chilli pepper for 10 minutes, stirring now and then. Increase the heat, add the meat and a pinch of salt and pepper, and cook for 2 minutes, stirring continuously. Add the finely chopped tomatoes, the vinegar, sultanas, sugar, cloves and cinnamon. Stir

well, turn down the heat and simmer, uncovered, over a very low heat for 15 minutes or until all excess moisture has evaporated, stirring at frequent intervals. Stir in the almonds, cook for a further 2–3 minutes and then remove from the heat.

Make the sauce. Put the walnuts and almonds in the blender with the cream, coriander leaves, cinnamon and a pinch each of salt and sugar; process briefly.

Fill the peeled peppers with the hot stuffing mixture, pressing it down into them gently but firmly.

Place each stuffed pepper on a plate, pour a little of the cream sauce over each serving and sprinkle with a few pomegranate seeds and a couple of coriander leaves (the colours echo those of the Mexican flag). Place a wedge of lime on each plate for each person to squeeze over the stuffed pepper.

Serve at once. If you prefer to serve the peppers very hot, return them to a moderate oven (325°F (170°C) mark 3) for 15 minutes in an ovenproof dish after stuffing them, to heat through before covering with the sauce and garnishing.

———— • ————

PEPPER, COURGETTE AND TOMATO FRICASSÉE
Pisto Manchego

3 large green peppers

½–1 green chilli pepper

3 medium-sized courgettes

3 medium-sized onions

10 medium-sized ripe tomatoes

pinch oregano

pinch basil

6 eggs, 2 of which hard-boiled

olive oil

salt and pepper

The Spanish equivalent of the French dish *pipérade*, this dish comes from the La Mancha region. Rinse, dry and prepare the peppers and the chilli pepper and chop them coarsely. Dice the courgettes. Chop the onions coarsely. Blanch, peel and seed the tomatoes and chop coarsely.

Heat 6 tbsp olive oil in a large, heavy-bottomed flameproof casserole dish and when it is very hot, add the peppers, chillis, courgettes and onions. Stir well, sprinkle with a pinch of salt and some freshly ground black pepper, cover and reduce the heat to very low. Simmer gently for 25 minutes.

Prepare the garnish. Cut the 2 hard-boiled eggs lengthwise in half, remove the yolks and rub these through a sieve; cut the whites into very small dice. Reserve this garnish.

Mix the chopped tomato flesh with a pinch of salt, 1 tbsp oil, the oregano and a small pinch of dried basil in a small saucepan; bring to the boil, stirring. Simmer, uncovered, for about 10 minutes, stirring frequently as the sauce reduces and thickens considerably. When it is very thick, stir into the pepper and onion mixture and cook over a low heat, stirring, for 1–2 minutes to reduce the liquid further. The vegetable mixture should be very thick, with no visible liquid. Beat 4 eggs in a bowl with a pinch of salt and pepper, stir into the vegetables and cook, stirring, over a very low heat. When the eggs have started to set and thicken the mixture, which will take about 30 seconds, remove from the heat and serve immediately, decorating each serving with a little of the egg garnish.

———— • ————

CATALAN LOBSTER WITH PEPPERS

1 large yellow pepper, roasted
1 large red pepper, roasted
2 frozen lobsters
4 large spring onions
4 medium-sized ripe tomatoes
10 black olives, stoned
10 green olives, stoned
generous pinch oregano
1½ tbsp capers
1 lemon
2 sprigs basil
4 tbsp light olive oil
2 tbsp extra-virgin olive oil
1 red chilli pepper, finely chopped (optional)
salt and black peppercorns

Roast the peppers as described on page 16, peel them and cut into thin strips. Cut these strips into small dice shortly before using. Thaw the lobsters very slowly, leaving them on the bottom shelf of the refrigerator for 30 hours or thaw more quickly at room temperature (about 3 hours). When completely thawed, poach in gently simmering salted water or in a *court-bouillon* for 5 minutes; drain and leave to cool. Place each one in turn on a wooden chopping board, the right way up, and place the tip of a heavy kitchen knife in the centre of the lobster's head where there is usually a cross-shaped indentation. Push firmly downwards or use a mallet to tap the handle of the knife sharply to pierce the shell, cut the lobster lengthwise in half and remove all the white meat. Reserve the soft meat in the head for other uses. Cut the white tail meat into slices about ¼ in (6–7 mm) thick and season.

Trim off the roots and leaves of the spring onions, remove the outermost layer and slice the bulb into thin rings; blanch these for 30 seconds in simmering salted water; remove with a slotted spoon and drain. Blanch, peel and seed the tomatoes; cut them into small pieces. Heat 2 tbsp of the light olive oil in a large non-stick frying pan and sauté the spring onions, chilli pepper, tomatoes, peppers and all the olives. Add the oregano, capers, a pinch of salt and some freshly ground black pepper. Reduce the heat and stir, simmering, for about 30 seconds. Turn off the heat. Heat 2 tbsp of light olive oil in another non-stick frying pan and sauté the pieces of lobster over a moderately high heat for about 30 seconds on each side. Spoon some of the vegetable mixture on to heated individual plates, place an equal number of lobster slices on top of each serving and squeeze a little lemon juice over them; sprinkle with olive oil and garnish with small sprigs of fresh basil or parsley.

———— • ————

AVOCADO AND CHICKEN SALAD WITH BLUE CHEESE DRESSING

3 ripe avocadoes

1 chicken, poached in aromatic stock (see page 72)

1 small bunch lamb's lettuce

1 head curly endive or frisée

1 bunch salad rocket

1 chicory

1 head radicchio

4 baby courgettes

2 cloves garlic, peeled and crushed

18 fl oz (500 ml) mayonnaise (see page 35)

2 oz (50 g) Gorgonzola, Roquefort or Stilton cheese

3½ fl oz (100 ml) single cream

2 tbsp finely chopped parsley

1 lemon

3 tbsp olive oil

salt and white peppercorns

Cook the chicken as described on page 72. While it is cooking, make the mayonnaise and set aside. Trim, wash and dry all the salad leaves and cut them into strips. Mix well and place on individual plates. Rinse, dry and trim the ends from the courgettes and slice them into thin rounds. Heat 3 tbsp olive oil in a wide non-stick frying pan and sauté the courgette slices with the 2 whole garlic cloves, slightly crushed with the flat of a large knife blade. Season with salt and pepper and add most of the parsley, reserving a little for a final garnish; when the slices are tender but still crisp, take them out of the pan,

leaving the garlic behind, and place on top of the shredded salad bed.

Stir the cream into the mayonnaise; add the blue cheese, rubbed through a fine sieve. Season with salt and freshly ground white pepper to taste and stir in a little lemon juice.

Slice the cooked chicken, cooled, into fairly small strips. Peel the avocadoes, cut lengthwise in half and remove the stone; cut lengthwise into thin slices and fan these out on top of the courgettes and salad. Top each serving with some of the chicken and cover with a few spoonfuls of the blue cheese dressing; decorate with sprigs of parsley and serve, handing round the remaining dressing in a bowl. Serves 6.

———— • ————

TEMPURA WITH TENTSUYU SAUCE

16 large courgette or pumpkin flowers (optional)

14 oz (400 g) prepared pumpkin

8 baby courgettes

16 thin spring onions

sunflower oil

1 egg yolk

4 oz (225 g) plain flour

salt

For the Tentsuyu sauce:

1 tsp crumbled instant dashi cube

4 fl oz (120 ml) Japanese light soy sauce

4 fl oz (120 ml) mirin (sweet Japanese rice wine) or dry sherry sweetened with 2 tsp caster sugar

momiji oroshi (grated daikon root mixed with finely chopped chilli pepper)

Serve with:

steamed or boiled rice

Make the batter: beat the egg yolk well until creamy in a large bowl with ½ tsp salt; gradually beat in 9 fl oz (250 ml) iced water, adding a little at a time. Sift in the flour and stir only just enough to mix; do not worry if there are a few small lumps (if stirred too much, this batter becomes gluey). Cover and chill in the coldest part of the refrigerator for at least 1 hour.

Prepare the vegetables, trimming, rinsing and drying them. Leave about 1 in (2½ cm) of stalk attached to the flowers, if using; remove the pistil from inside the flower carefully and then rinse the flowers; spread out on a clean cloth to dry. Cut the peeled pumpkin into pieces just under ¼ in (½ cm) thick and about 2 in (5 cm) long. Cut the courgettes lengthwise into slices just under ¼ in (½ cm) thick and then cut each of these long slices across, in half. Trim the roots and the ends of the leaves off the spring onions, remove the outermost layer of the bulb and leaves and cut the inner, tender part of both lengthwise in half.

Make the sauce: pour 9 fl oz (250 ml) water into a small saucepan, add the crumbled instant dashi cube, the soy sauce and the mirin; stir well as you heat to just below boiling point. Turn off the heat; this will be reheated, without letting it boil, just before serving. Make the *momiji oroshi*, ready to be added to the sauce just before serving: peel one-third of a daikon root, cut into round slices and process in the food processor with 1–2 chilli peppers (seeds and stalks removed) until it forms a thick purée. Place this purée in a fine mesh plastic sieve and leave to drain over a bowl.

Cook the rice. When the rice is ready, reheat the sauce and pour into tiny bowls, one for each person; stir about 1 tbsp of the *momiji oroshi* into each bowl. Serve the rice in a large bowl for people to help themselves or in small, individual bowls. Fry the vegetables just before serving. Heat plenty of oil in a deep-fryer to 350°F (180°C). Divide the prepared vegetables into 4 batches; you will serve each batch as soon as it is fried. Stir the chilled batter very briefly and dip each piece of vegetable in it before adding to the hot oil, working quickly once the first battered piece has been added to the oil. At this temperature the batter-coated vegetables will fry gently and will take a few minutes to become lightly crisp and pale golden brown; as soon as they do so, take out of the oil, drain briefly on kitchen paper and

serve at once. While your guests or family are eating this first batch, fry the next and so on.

To eat, pick up a piece of fried vegetable with chopsticks and dip it in the sauce. The rice makes a suitably bland foil; if there is any sauce left over, it can be poured over the remaining rice.

———•———

HUNGARIAN COURGETTES WITH DILL SAUCE

6 large courgettes
1 medium-sized onion, finely chopped
10 oz (300 g) minced lean veal
3½–4 oz (100 g) minced cooked ham
1 egg
2 oz (50 g) long-grain rice (uncooked weight), boiled or steamed
1½ oz (40 g) grated Gruyère cheese
4 tbsp chopped fresh dill
10 fl oz (300 ml) single cream
1 tbsp plain flour
1 lemon
1 tbsp sunflower oil
2 oz (50 g) butter
white wine vinegar
nutmeg
salt and pepper
Serve with:
Sweet-sour Baby Onions (see page 192)

Preheat the oven to 350°F (180°C) mark 4. Bring a large saucepan two-thirds full of salted water to the boil; add 2 tbsp vinegar. Rinse the courgettes and cut off about ¾ in (2 cm) from each end. Use a potato peeler to remove a very thin layer of skin from them;

if they are very long cut them in half. Hollow out the centre of each courgette with an apple corer. Blanch in the boiled acidulated water for 5 minutes, drain and leave to cool. Fry the onion very gently for 10 minutes in 1 oz (25 g) butter and 1 tbsp oil, stirring frequently. When tender but not browned, stir in the meat, season with a little salt and pepper and stir over a higher heat for 2 minutes. Transfer to a bowl and mix with the minced ham, the egg, cheese and boiled rice, adding a little more salt and pepper to taste and a pinch of nutmeg. Stuff the hollowed courgettes with this mixture. Oil a wide, shallow ovenproof dish and arrange the stuffed courgettes in it in a single layer.

Melt 1 oz (25 g) butter in a small saucepan and add the dill; fry gently for 1 minute; add 3½ fl oz (100 ml) water and the finely grated rind of the lemon. Simmer, uncovered, for 2 minutes. Mix the cream with the flour and stir into the contents of the saucepan, continuing to stir and cook until the sauce has thickened. Stir in the lemon juice and season with salt and pepper. Pour this sauce all over the stuffed vegetables. Bake in the oven for 15–20 minutes, then serve without delay. Serves 6.

———•———

PERSIAN STUFFED COURGETTES
Dolmeh Kadu

6 large courgettes
1 cup dried split green peas
1 large onion, finely chopped
1 lb (450 g) minced lean lamb or beef
generous pinch ground cinnamon
2 oz (50 g) butter
salt and pepper
Serve with:
natural yoghurt, lightly salted to taste

Rinse the courgettes and prepare as for the previous

recipe, cutting off their ends, hollowing out the centres and blanching them. Rinse the dried peas well in a sieve under running cold water; boil in salted water for 30 minutes or until tender. Drain and reserve.

Sweat the onion in the butter for 10 minutes over a very low heat, stirring frequently; turn up the heat and add the meat, sprinkle with a generous pinch of cinnamon and cook for 2 minutes while stirring continuously over a moderately high heat. Season with salt and pepper and leave to cool. Mix with the peas, adding a little more salt, pepper and cinnamon to taste.

Stuff the hollowed courgettes with this mixture (see previous recipe), arrange in a single layer in a wide, shallow flameproof dish, add 18 fl oz (500 ml) water, cover with a lid or with foil and simmer very gently for 25–30 minutes or until the vegetables are tender. Transfer to heated individual plates, handing round a bowl of natural yoghurt mixed with a pinch of salt. Serves 6.

———•———

COURGETTE FRITTERS
Mücver

2¼–2½ lb (1 kg) courgettes
1 medium-sized onion
2 egg yolks
7 oz (200 g) plain flour
7 oz (200 g) feta cheese, crumbled
2 tbsp chopped fresh dill leaves
2 tbsp chopped parsley
olive oil
nutmeg
salt and pepper
Garnish with:
sprigs of flat-leaved parsley or dill
Serve with:
Tsatsiki (see page 149)

Trim off the ends of the courgettes, peel them with the potato peeler and grate finely into a large mixing bowl. Do likewise with the onions. Stir in the sifted flour, the finely crumbled feta cheese and the lightly beaten eggs. Season with a very little salt if wished, with plenty of freshly ground pepper and a pinch of nutmeg. Stir in the dill and parsley, blending the mixture very thoroughly.

Heat enough oil for shallow-frying in a wide frying pan or wok. When hot, drop tablespoonfuls of the mixture into the hot oil, taking care to space them out so that they do not stick to one another; fry until crisp and golden brown, turning once. Remove with a slotted spoon and finish draining on kitchen paper.

Serve garnished with sprigs of parsley or dill, with tsatsiki.

BRAISED ARTICHOKES

8 very young tender globe artichokes
1 lemon
¼ chicken stock cube
1 large clove garlic, peeled
1 tbsp chopped parsley
4 tbsp extra-virgin olive oil
salt and pepper

Prepare the artichokes, removing the outer leaves and chopping the top third off the remaining, tender heart; trim the stems to about 2 in (5 cm) in length and use a potato peeler to peel off the tough outer layer (see page 18). The artichokes most suitable for use are the very tender varieties; if you grow your own Breton variety, use side buds, trimmed off to let the main bud develop; they will not be old enough to have developed a choke. Rinse them inside and out. As each one is cut, peeled and trimmed, drop it into a large bowl of water that has been acidulated with the lemon juice. Drain the artichokes, place them in one layer, turned upside down in a saucepan that is just wide enough to ac-

commodate them all. Add 10 fl oz (300 ml) water, the oil, the crumbled stock cube, garlic, parsley and the pieces of stalk you have trimmed off, also peeled. Add a very small pinch of salt and some freshly ground pepper.

Bring to the boil over a moderate heat, cover tightly and turn down the heat to low; simmer gently for 15 minutes or until the artichokes are tender and the liquid has reduced considerably.

ARTICHOKE BOTTOMS IN CHEESE SAUCE

6 large globe artichokes
vegetable stock (see page 38)
juice of 1 lemon
1 oz (30 g) grated Gruyère cheese
butter
For the cheese sauce:
14 fl oz (400 ml) béchamel sauce (see page 36) or velouté sauce (see page 35)
2 egg yolks
1½ oz (40 g) grated Gruyère cheese
Garnish with:
tomato rosebuds (see page 27)
Serve with:
Courgette Purée (see page 150)

Gently boil the artichoke bottoms in vegetable stock. Fill a large bowl with water acidulated with the lemon juice. Prepare the artichokes as described on page 18, stripping each one right down to its fleshy base; remove the choke and cut off the stalk. Use a small sharp knife to pare away any rough edges and drop immediately into the acidulated water. When they are all ready, poach in the cooking stock for about 20 minutes or until they are tender. Preheat the oven to 350°F (180°C) mark 4. Make the cheese sauce: make the béchamel or velouté sauce, which should be of pouring consistency and not too thick for this recipe. As soon as it has thickened, take the saucepan off the heat and beat with a balloon whisk as you add the egg yolks one at a time, followed by the grated cheese.

Drain the cooked artichoke bottoms and cut each one into 3 or 4 thin sections. Cover the bottom of the dish with half the sauce, place the sliced artichokes on top, cover with the remaining sauce and sprinkle with the grated cheese. Dot little flakes of butter here and there on the surface. Bake in the oven for 15 minutes, to form a golden brown layer on top. Serve immediately, garnished with a few tomato rosebuds.

Courgette purée makes a good complement to the artichokes, both in taste and appearance.

---•---

CRISPY GLOBE ARTICHOKES

8 baby globe artichokes

juice of 1 lemon

light olive oil

salt

If you grow your own artichokes, the small side buds can be used for this recipe. Large Breton artichokes are too tough. Half fill a very large saucepan with salted water, add the lemon juice and bring to the boil. Prepare the artichokes, trimming off the stems about 2 in (5 cm) from their bases; scrape off the outer layer from the stem; remove the lower, outer leaves. Remove the feathery choke enclosed by the leaves. Rinse thoroughly, drain well and place upside down on a chopping board. Take hold of the stems one at a time and rotate the artichokes, pressing down hard against the board; to make them open out like flowers.

Blanch the artichokes for a full 5 minutes in the fast boiling acidulated water, drain well and place them upside down on a chopping board to dry off completely. Pour sufficient olive oil into a very wide, heavy-bottomed saucepan to give a depth of about

2 in (5 cm). When it is very hot, but not smoking, place the artichokes upside down in it and fry over a moderately high heat until they are tender and the leaves are crisp and lightly browned. If using a deep-fryer, the oil should be kept at an even 350°F (180°C) throughout the frying time. Serve hot.

---•---

ARTICHOKE BOTTOMS WITH GARLIC

12 large globe artichokes

1 lemon

7 fl oz (200 ml) chicken stock (see page 37)

4 tbsp olive oil

2 cloves garlic, peeled

2 tbsp chopped parsley

salt and white peppercorns

Prepare the artichokes in the same way as for Artichoke Bottoms in Cheese Sauce on page 144. Place them in a bowl of acidulated water.

Choose a very large frying pan or saucepan to accommodate all the drained artichoke bottoms in a single layer (or use 2 pans). Pour the stock and the oil all over them. Add the whole but lightly crushed garlic cloves and sprinkle with the parsley. Season with a small pinch of salt and some freshly ground pepper. Cover tightly and simmer for about 20 minutes, or until tender. Serve hot.

---•---

BAKED TOMATOES WITH MINT

2¼–2½ lb (1 kg) beef tomatoes

2 sprigs mint

extra-virgin olive oil

salt and pepper

Preheat the oven to 350°F (180°C) mark 4. Rinse the tomatoes and cut them horizontally in half. Lightly oil 2 wide roasting tins and place the halved tomatoes, cut side uppermost, in a single layer in them. Season with salt and freshly ground pepper and sprinkle with a little olive oil.

Tear the mint leaves into small pieces and sprinkle these over the cut surfaces of the tomatoes. Pour in just enough water, between the tomatoes, to cover the bottom of the tins and come about one-quarter of the way up the sides of the tomatoes. Bake for about 50 minutes.

———•———

TOMATOES AU GRATIN

4 beef tomatoes
1½ tbsp chopped parsley
1 small clove garlic, peeled
1 tbsp capers
6 tbsp fine breadcrumbs
oregano
extra-virgin olive oil
salt and pepper

Preheat the oven to 320°F (160°C) mark 2½. Chop the parsley very finely with the garlic and transfer to a bowl. Chop the capers coarsely and place in the bowl; add 3 tbsp of the breadcrumbs and mix well. Stir in 1 tbsp oil and a generous pinch each of oregano and freshly ground black pepper.

Rinse the tomatoes and cut horizontally in half. Place cut side uppermost on an oiled baking sheet. Sprinkle with a little salt; cover each cut surface with some of the prepared mixture. Sprinkle the topping on each tomato with a little oil and cover with a layer of plain breadcrumbs. Bake in the oven for 25–30 minutes, placing the tomatoes under a very hot grill for the last 3 minutes to brown the topping. Serve hot or warm.

BAKED AUBERGINES

4 medium-sized long aubergines
2 cloves garlic, peeled
extra-virgin olive oil
salt and pepper

Preheat the oven to 350°F (180°C) mark 4. Trim off the stalks, cut the aubergines lengthwise in half and make deep intersecting cuts in the exposed flesh to form a lattice design. Season with salt and freshly ground pepper. Sprinkle with the finely chopped garlic, pressing it down into the deep cuts in the flesh.

Oil a roasting tin and arrange the aubergines, cut sides uppermost, in a single layer. Sprinkle the cut surfaces with olive oil and bake for about 35 minutes or until the aubergines are tender and are lightly browned on the surface. Serve hot.

———•———

AUBERGINE SALAD

2¼ lb (1 kg) long firm aubergines
2–3 cloves garlic, peeled and thinly sliced
1 red chilli pepper, fresh or dried, seeds removed
generous pinch oregano
3 tbsp wine vinegar
extra-virgin olive oil
salt and pepper
Serve with:
fresh, soft or mild, sliceable cheese
crusty white bread or wholemeal bread

Bring a large saucepan of salted water to the boil. While it is heating, chop off the ends of the aubergines, rinse and dry them and cut lengthwise in quarters.

Slice the flesh lengthwise into strips, cutting these in half if they are very long. Soak in a large bowl of lightly salted water for about 10 minutes or until the water turns a brownish-grey colour. Drain the aubergine strips and blanch in the saucepan of boiling water for 1½ minutes. Drain again and spread out in a single layer on a clean cloth; cover with another cloth and blot dry.

Place the aubergine strips in a wide, shallow serving dish, sprinkle with the very thinly sliced garlic, the finely chopped or crumbled chilli pepper, oregano and a little salt and pepper. Sprinkle with the vinegar and about 4 tbsp olive oil. Stir and leave to marinate for at least 30 minutes before serving.

———— • ————

AUBERGINE AND MINT RAITA

1¼ lb (600 g) white or purple aubergines
1 large spring onion
generous 1 pint (600 ml) low-fat natural or Greek yoghurt
1 large sprig mint
salt and pepper
Garnish with:
small mint leaves
Serve with:
curry

Set up the steamer ready for cooking the vegetables. Trim and peel the aubergines, cut the flesh into ¾-in (2-cm) cubes and steam for 10 minutes. Remove the root end and outermost layer of the spring onion; cut the inner, tender parts of the bulb and leaves into very thin rings. Blanch these for 5 seconds in boiling salted water and drain.

Mix the spring onion with the yoghurt in a large bowl, adding ¾ tsp salt and plenty of freshly ground pepper. Mix very thoroughly. Stir in 1 tbsp chopped mint.

Place the cooked aubergine cubes in a large bowl, crush or mash them coarsely with a fork or potato masher and leave to cook. Mix them with the yoghurt, add a little more salt and pepper if needed and transfer to a serving dish. Decorate with small mint leaves. This cooling raita provides a refreshing foil to curried dishes.

———— • ————

DEEP-FRIED AUBERGINE WITH MISO SAUCE
Nasu no Miso Kake

4 very small white or purple aubergines
sunflower oil
salt
For the miso sauce:
2 tbsp shiro miso (sweet white bean paste)
4 tbsp saké, dry sherry or dry vermouth
2 tsp caster sugar
½ tsp cornflour or potato flour
generous pinch monosodium glutamate (optional)
Garnish with:
radish flowers (page 28)
parsley sprigs

Slice off the ends of the aubergines; rinse and dry them and cut lengthwise into quarters; cut across the middle of each quarter to obtain pieces about 2½ in (6 cm) long. Use a small, sharp vegetable knife to 'turn' these pieces, cutting off the sharp edges. Make 2 or 3 parallel lengthwise cuts, about ⅛ in (3 mm) deep in the skin of each piece and place in a large bowl of cold, salted water; leave to stand. The water should cover all the pieces. Drain after 20 minutes and blot-dry in clean cloths.

Make the miso sauce: warm the sweet white bean paste gently with the saké and sugar, stirring with a wooden spoon. Mix the cornflour or potato flour with 4 tbsp cold water and stir into the mixture; keep stirring as the sauce comes to the boil and thickens. Quickly draw aside from the heat and stir in the monosodium glutamate, if used. Do not add salt.

Heat plenty of sunflower oil in a deep-fryer to 320°F (160°C) (or use a wok); fry the pieces of aubergine in batches until they are pale golden brown but not crisp; each batch will take about 5–7 minutes. Remove with a slotted spoon and finish draining on kitchen paper while keeping hot. When the last batch is cooked, serve without delay on individual plates, coated with the sauce. Garnish with radish flowers and sprigs of parsley. Serves 6.

— • —

FRIED AUBERGINES WITH GARLIC AND BASIL

2 large firm aubergines
2 large cloves garlic, peeled
4 tbsp olive oil
salt and pepper
Garnish with:
basil leaves

Trim off the ends of the aubergines; peel and quarter them lengthwise. Scoop out the central section if there are seeds in it. Chop into dice. Heat the oil in a very large, non-stick frying pan and stir-fry the aubergine dice over a fairly high heat. As soon as they have browned lightly, cover the pan, reduce the heat to very low and cook gently for 15 minutes; they will produce some moisture but you may need to add 2–3 tbsp water after a while. Season with salt and pepper and serve, garnished with fresh basil leaves.

— • —

PEPPERS AU GRATIN

5 large peppers (red, yellow, green and purple)
5 cloves garlic, peeled
2 tinned anchovy fillets, drained
12 fleshy black olives
1½–2 tbsp capers
few basil leaves
generous pinch oregano
fine dry breadcrumbs
extra-virgin olive oil
salt and pepper

Preheat the oven to 400°F (200°C) mark 6. Rinse and dry the peppers, wrap each one tightly in foil, enclosing it completely, and bake in the oven for 40 minutes, turning halfway through this time. Take the parcels out of the oven and leave to cool completely (chill in the refrigerator for 1 hour if you wish) before unwrapping them. Nearly all the skin will come away with the wrapping; what little is left will peel off easily. Remove and discard the stalk and seeds; cut the flesh into thin strips.

Oil two wide, shallow ovenproof dishes and sprinkle the basil leaves, torn into very small pieces, over the bottom. Use parsley instead if preferred. Spread half the pepper strips out in a single layer in each dish, mixing the colours. Season with salt and pepper. Chop the garlic and the anchovies fairly coarsely and sprinkle over the peppers. Stone the olives, cut them in quarters and sprinkle evenly over the peppers. Finish with a pinch of oregano for each dish and a thick covering of breadcrumbs. Drizzle a little olive oil over this topping and bake in the oven for 15 minutes to form a crunchy, golden brown layer on the surface.

— • —

STIR-FRIED PEPPERS IN SHARP SAUCE

2¼ lb (1 kg) peppers

4 cloves garlic, peeled

1 chilli pepper, fresh or dried

3½ fl oz (100 ml) wine vinegar

6 tbsp fine breadcrumbs

1½–2 tbsp capers

oregano

4 tbsp extra-virgin olive oil

salt

Prepare the peppers: remove the stalks, seeds and white membrane. Cut them into fairly thin strips. Heat the oil in a very large frying pan or wok and stir-fry the garlic, chilli pepper and pepper strips for 2–3 minutes over a high heat. Reduce the heat and sprinkle with about one-third of the vinegar and 2 tbsp of the breadcrumbs. Add a pinch of salt. Keep stirring the peppers as you add the remaining vinegar and breadcrumbs a little at a time. Stir in the capers. Cover and cook over a low heat for about 15 minutes, or until the peppers are tender, moistening with a little water when necessary. Add a little more salt if needed, a pinch of oregano, stir once more and remove from the heat.

This mixture also makes a delicious stuffing for vegetables and poultry.

———— • ————

TSATSIKI

1 medium-sized cucumber

1 large clove garlic, peeled and finely grated

18 fl oz (500 ml) natural yoghurt

2 tbsp chopped fresh dill leaves

few fleshy black olives

olive oil

salt

Serve with:

Courgette Fritters (see page 143)

Wash and dry the cucumber, slice off the ends and peel. Grate into a large bowl, mix in the garlic, ½ tsp salt, the dill and the yoghurt. Drizzle a little oil over the surface, garnish with a few black olives and serve lightly chilled as a refreshing side dish to a Greek or Turkish meal.

———— • ————

STIR-FRIED COURGETTES WITH PARSLEY, MINT AND GARLIC

1 lb (500 g) baby courgettes

2 tbsp chopped parsley

1 sprig mint

2 cloves garlic, peeled

4 tbsp olive oil

wine vinegar

salt and pepper

Chop the parsley and tear the mint leaves into small pieces by hand. Rinse and dry the courgettes, slice off the ends and cut them lengthwise into ¼-in (6–7-mm) thick slices. Cut these into 1½-in (4-cm) lengths.

Heat the oil in a large frying pan and stir-fry the pieces of courgette with the slightly crushed but whole garlic clove over a fairly high heat for approx. 5 minutes; they should be just tender but still crisp. Remove the garlic clove. Sprinkle the courgettes with 3 tbsp vinegar, a pinch of salt and some freshly ground pepper. Stir-fry for a few seconds, reduce the heat and then sprinkle over the parsley and mint, stir briefly and then remove from the heat. Drain off any excess oil and serve the courgettes at

149

once. This dish goes well with grilled meat, poultry or fish. It is also excellent served cold with cured meat, sausages and smoked cheese.

———•———

COURGETTE PURÉE

2¼–2½ lb (1 kg) courgettes

14 fl oz (400 ml) vegetable stock (see page 38) or chicken stock (see page 37)

1 clove garlic, peeled

4 egg yolks

4 tbsp finely grated hard cheese

few basil leaves

nutmeg

1½ oz (40 g) unsalted butter

salt and white peppercorns

Garnish with:

sprigs of basil

Rinse and dry the courgettes and slice them into thick rounds. Place these in a saucepan with the stock and the garlic clove, partially crushed with the flat of a heavy knife blade. Cover and bring to a gentle boil, then simmer for about 20 minutes or until the courgettes are tender but still firm. Drain, discard the garlic clove and put the courgettes through a vegetable mill together with a few basil leaves torn into small pieces. (You may prefer to reduce the vegetables to a thick purée by beating with a hand-held electric beater.)

Beat the egg yolks in a large bowl with the grated cheese, a pinch of salt, some freshly ground white pepper and a pinch of nutmeg. Heat the purée, add the butter and beat as it melts; keep beating as you add the egg yolk and cheese mixture a little at a time. Do not allow the purée to boil. Stir over a low heat until it is very hot. Garnish with basil.

COURGETTES RUSSIAN STYLE

1¾ lb (800 g) baby courgettes

5 fl oz (150 ml) single cream

1 tbsp chopped parsley

1 tbsp chopped mint

plain flour

2 oz (50 g) butter

2 tbsp sunflower oil

salt and pepper

Trim off the ends of the courgettes. Wash and dry them and cut into thick rounds. Place these in a bowl, dust with flour and coat all the pieces, turning them over and mixing by hand. Tap the bottom of the bowl on the work surface to make any excess flour fall to the bottom. Heat the butter with the oil in a large frying pan and when very hot add the courgette rounds, leaving the unwanted flour behind in the bowl; fry over a moderately high heat for 2–3 minutes, stirring continuously. Reduce the heat to low, cover and cook gently for about 10 minutes, stirring occasionally, until the courgettes are tender but still crisp. Season with a little salt and pepper, pour in the cream, add the parsley and mint and continue cooking while stirring gently over a low heat for 1–2 minutes. Serve very hot.

———•———

AVOCADO CREAM DESSERT

2 large ripe avocadoes

1 large lime

5 tbsp icing sugar, sifted

Decorate with:

small mint leaves

thin lime slices

Cut the avocadoes lengthwise in half, remove the stones and peel. Cut the pulp into fairly small pieces and purée in a food processor or put through a vegetable mill with a fine gauge disc fitted. Immediately mix this purée with part or all of the lime juice to prevent discoloration and sweeten with the icing sugar. Stir well. Spoon the avocado cream into ice cream coupes or crystal dishes, cover with cling film and chill. The dessert can be made up to 1 hour in advance; it is best served chilled but not ice cold. Decorate with mint leaves and lime slices.

———•———

AVOCADO ICE CREAM

2¼ lb (1 kg) avocado flesh (net weight)

3 tbsp liquid glucose or acacia honey

1 lb 10 oz (750 g) caster sugar

juice of 2 freshly pressed limes

Decorate with:

clusters of redcurrants, or wild strawberries

Combine the liquid glucose and the sugar with 1¾ pints (1 litre) cold water in a saucepan. Bring to the boil, stirring frequently. When the sugar has completely dissolved, remove from the heat and leave to cool. Wait until the syrup is cold before you start preparing the avocadoes (their flesh discolours very quickly once exposed to the air). Cut them in half, take out the stone and scoop all the flesh out of the skin. Working quickly, measure the required weight of flesh, place in the blender with the lime juice and 4 fl oz (120 ml) of the syrup; blend until very smooth. The lime juice prevents the flesh from discolouring.

Add the remaining syrup and blend. (If you do not have a large-capacity blender, you can process the mixture in two batches; stir the two batches thoroughly in a bowl before freezing). If you have an electric ice-cream maker, pour the mixture into it and process for 15 minutes. Transfer into suitable containers and place in the freezer for at least 2 hours. Alternatively, pour into ice trays and freeze these in the ice box of your refrigerator, stirring several times as soon as the mixture starts to freeze and thicken.

Ten minutes before serving the ice cream, take it out of the freezer. Use an ice cream scoop and transfer to glass bowls or coupes. Decorate each individual serving with redcurrants or wild strawberries.

———•———

CANDIED PUMPKIN

1 lb (500 g) pumpkin flesh (net weight)

1 lb (500 g) caster sugar

1 lemon

Peel the piece of pumpkin, remove the seeds and surrounding fibres and measure out the above weight. Cut into small cubes, place in a bowl and cover with cold water. Leave to soak for 24 hours, then drain.

Melt the sugar in 18 fl oz (500 ml) water over a low heat, stirring continuously; continue cooking over a low heat until the syrup temperature reaches about 244°F (118°C): use a sugar thermometer to measure the temperature if you have one, or allow a large drop to fall into a saucer of iced water; if it forms a soft ball, it is ready. Add the pumpkin, the juice of the lemon and its finely grated rind. Stir, cover the saucepan and cook gently for about 30 minutes or until the pumpkin is tender.

Leave to stand for one week at room temperature, then take out the pumpkin pieces and spread out in one or more very large sieves. Place in the oven at about 225°F (110°C) mark ¼ to dry out gradually; when dry, store in tightly sealed glass jars.

Excellent eaten as confectionery or with liqueurs at the end of a meal.

PUMPKIN AND AMARETTI CAKE

1 lb 2 oz (500 g) pumpkin flesh (net weight)

18 fl oz (500 ml) milk

5 oz (150 g) slightly stale white bread, crusts removed

3 eggs

2 oz (50 g) caster sugar

3½ oz (100 g) sultanas

3½ oz (100 g) pine nuts

9 double Amaretti di Saronno (i.e. 18 small biscuits)

softened butter

flour

coarse sea salt

Dice the pumpkin flesh and cook in the pressure cooker with 7 fl oz (200 ml) water and a pinch of coarse sea salt for 20 minutes.

Preheat the oven to 350°F (180°C) mark 4. Prepare all the other ingredients: bring the milk to the boil in a saucepan, draw aside from the heat and place the bread to soak in it. After 10 minutes, remove the bread from the milk with a slotted spoon. Do not squeeze the milk out of the bread but beat it with a fork until smooth. The excess milk left behind in the saucepan can be poured away. When the pumpkin is cooked, reduce it to a coarse purée by crushing it with a fork or a potato masher. Mix very thoroughly with the bread purée.

Beat the eggs lightly with the sugar and stir into the pumpkin mixture, adding the sultanas, pine nuts and all but one of the Amaretti biscuits (i.e. half the contents of one of the twists of tissue paper), coarsely crushed. Grease the inside of a 9½-in (24-cm) deep tart tin with butter, sprinkle with a little flour, tipping out excess and turn the mixture into this tin; smooth the surface level with a palette knife. Bake in the oven for a total of 30 minutes; after 20 minutes, open the oven door and quickly sprinkle the remaining Amaretto, very coarsely crumbled, on to the centre of the cake and close the oven door again for the last 10 minutes of the baking time. Serve warm.

TURKISH PUMPKIN DESSERT
Kabak Tatlisi

2¼ lb (1 kg) pumpkin flesh

5 oz (150 g) caster sugar

5 oz (150 g) chopped walnuts

Cut the pumpkin flesh into thin, rectangular pieces measuring about 3 × 1¼ in (8 × 3 cm). Place these in a very wide, flameproof earthenware casserole dish, sprinkle with the sugar and add just sufficient water to cover; leave to stand for 24 hours. Place the casserole dish on a moderate heat, bring to the boil, reduce the heat and simmer gently for 45 minutes, stirring now and then, until the pumpkin is tender. Leave to cool at room temperature and then serve in small bowls or dishes, sprinkled with the chopped walnuts.

·BULB AND ROOT VEGETABLES·

CARROT BREAD

Babka

p. 161

Preparation: 30 minutes + 1 hour 30 minutes rising time
Cooking time: 50 minutes
Difficulty: easy
Appetizer

ONION SALAD WITH COLD MEAT

p. 161

Preparation: 5 minutes
Difficulty: easy
Appetizer

ONIONS WITH DRESSING

p. 161

Preparation: 40 minutes
Cooking time: 20 minutes
Difficulty: easy
Appetizer

CARROT SALAD TURKISH STYLE

Yogurtlu Havuç Salatasi

p. 162

Preparation: 20 minutes
Difficulty: very easy
Appetizer

CRUDITÉS WITH SESAME DIP

p. 162

Preparation: 30 minutes
Difficulty: very easy
Appetizer

CARROT, CELERY AND FENNEL SALAD WITH PRAWNS

p. 163

Preparation: 45 minutes
Cooking time: 5 minutes
Difficulty: easy
Appetizer

DAIKON ROOT SALAD WITH CHILLI AND SALMON ROE

Ikura to daikon no oroshi ae

p. 163

Preparation: 10 minutes
Difficulty: very easy
Appetizer

RADISH AND CRESS SALAD IN CREAM DRESSING

p. 163

Preparation: 15 minutes
Difficulty: very easy
Appetizer

CELERIAC SALAD

p. 164

Preparation: 25 minutes
Difficulty: easy
Appetizer

RUSSIAN BEETROOT PÂTÉ

p. 164

Preparation: 25 minutes + chilling time
Cooking time: 27 minutes
Difficulty: very easy
Appetizer

BEETROOT AND CARDOON SALAD

p. 164

Preparation: 25 minutes
Cooking time: 1 hour 30 minutes–2 hours
Difficulty: very easy
Appetizer

BAKED SWEET POTATOES WITH CREAM DRESSING

p. 165

Preparation: 15 minutes
Cooking time: 45 minutes
Difficulty: very easy
Appetizer

Pasta and potatoes

TURKISH POTATO SALAD

Patates Salatasi

p. 165

Preparation: 15 minutes
Cooking time: 30 minutes
Difficulty: very easy
Appetizer

BAKED POTATOES WITH CAVIAR AND SOURED CREAM

p. 166

Preparation: 10 minutes
Cooking time: 50 minutes
Difficulty: very easy
Appetizer

HUNGARIAN POTATO SALAD

p. 166

Preparation: 20 minutes
Cooking time: 20 minutes
Difficulty: very easy
Appetizer

JERUSALEM ARTICHOKE SALAD

p. 166

Preparation: 30 minutes
Difficulty: easy
Appetizer

RUSSIAN CARROT AND GREEN APPLE SALAD

p. 167

Preparation: 30 minutes
Difficulty: very easy
Appetizer or accompaniment

GRATED CARROT SALAD WITH INDIAN MUSTARD DRESSING

p. 167

Preparation: 25 minutes + 30 minutes soaking time
Difficulty: very easy
Appetizer or accompaniment

CARROT AND ONION SALAD WITH LIME AND GINGER DRESSING

p. 167

Preparation: 25 minutes
Difficulty: very easy
Appetizer or accompaniment

JAPANESE DAIKON ROOT AND CABBAGE SALAD

Namasu

p. 168

Preparation: 25 minutes
Difficulty: easy
Appetizer or accompaniment

JAPANESE MIXED SALAD WITH SAMBAI ZU DRESSING

p. 168

Preparation: 20 minutes
Difficulty: easy
Appetizer or accompaniment

POTATO SALAD WITH SPRING ONION AND CHIVES

p. 169

Preparation: 20 minutes
Cooking time: 20 minutes
Difficulty: very easy
Appetizer or accompaniment

CHILLED LEEK AND POTATO SOUP

Vichyssoise

p. 169

Preparation: 15 minutes + 2 hours chilling time
Cooking time: 50 minutes
Difficulty: very easy
First course

CREAM OF CELERIAC SOUP

p. 169

Preparation: 20 minutes
Cooking time: 40 minutes
Difficulty: very easy
First course

CREAM OF PARSNIP SOUP WITH BACON AND GARLIC CROUTONS

p. 170

Preparation: 25 minutes
Cooking time: 20 minutes
Difficulty: easy
First course

PHILADELPHIA PEPPER POT

p. 170

Preparation: 25 minutes
Cooking time: 1 hour 15 minutes
Difficulty: very easy
First course

RUSSIAN BEETROOT AND CABBAGE SOUP

Borscht

p. 171

Preparation: 30 minutes
Cooking time: 2 hours 20 minutes
Difficulty: easy
First course

ICED BEETROOT SOUP

p. 171

Preparation: 15 minutes + chilling time
Cooking time: 20 minutes
Difficulty: very easy
First course

CREAM OF POTATO AND LEEK SOUP

Potage Parmentier

p. 172

Preparation: 20 minutes
Cooking time: 50 minutes
Difficulty: very easy
First course

CREAM OF LEEK AND COURGETTE SOUP

p. 172

Preparation: 15 minutes
Cooking time: 50 minutes
Difficulty: very easy
First course

CREAM OF POTATO SOUP WITH CRISPY LEEKS

p. 173

Preparation: 30 minutes
Cooking time: 35 minutes
Difficulty: easy
First course

ONION SOUP

p. 173

Preparation: 30 minutes
Cooking time: 2 hours
Difficulty: easy
First course

ONION SOUP AU GRATIN

p. 174

Preparation: 35 minutes
Cooking time: 2 hours 30 minutes
Difficulty: easy
First course

PROVENÇAL GARLIC SOUP

Aïgo Bouïdo

p. 174

Preparation: 25 minutes
Cooking time: 35 minutes
Difficulty: easy
First course

SALSIFY AND MONKFISH SOUP

p. 174

Preparation: 25 minutes
Cooking time: 40 minutes
Difficulty: easy
First course

ICED HAWAIIAN SOUP

p. 175

Preparation: 25 minutes
Cooking time: 35 minutes
Difficulty: very easy
First course

SPAGHETTI WITH SPRING ONION, TOMATO AND HERB SAUCE

p. 175

Preparation: 35 minutes
Cooking time: 25 minutes
Difficulty: easy
First course

PASTA AND POTATOES

p. 176

Preparation: 20 minutes
Cooking time: 30 minutes
Difficulty: very easy
First course

PISSALADIÈRE

p. 176

Preparation: 20 minutes
Cooking time: 40 minutes
Difficulty: easy
First course

POTATO DUMPLINGS WITH BASIL SAUCE

Gnocchi al pesto

p. 177

Preparation: 45 minutes
Cooking time: 40 minutes
Difficulty: fairly easy
First course

GENOESE SEAFOOD AND VEGETABLE SALAD

Cappon magro

p. 177

Preparation + cooking time: 2 hours 30 minutes
Difficulty: easy
Main course

JERUSALEM ARTICHOKE VOL-AU-VENTS RUSSIAN STYLE

p. 179

Preparation: 30 minutes
Cooking time: 20 minutes
Difficulty: easy
Main course or appetizer

SPANISH OMELETTE

Tortilla de papas

p. 179

Preparation: 20 minutes
Cooking time: 20 minutes
Difficulty: easy
Main course

ONION AND BACON PANCAKES

p. 180

Preparation: 30 minutes + 6 hours chilling time
Cooking time: 40 minutes
Difficulty: fairly easy
Main course

Sweet-sour baby onions

159

Carrots with cream and herbs

160

CARROT BREAD
Babka

14 oz (400 g) carrots
scant 1 oz (20 g) fresh yeast or ½ oz (10 g) dried yeast
1 lb 2 oz (500 g) strong white flour
3 eggs (at room temperature)
3 oz (80 g) butter, melted
3½ oz (100 g) caster sugar
2½ oz (70 g) flaked or slivered almonds
finely grated rind of 1 lemon
vanilla essence
salt
Serve with:
soured cream
Russian Beetroot Paté (see page 164)

Peel the carrots and grate very finely. Crumble the yeast into a small bowl containing 3½ fl oz (100 ml) lukewarm water. Sift half the flour (9 oz (250 g)) into a large bowl; make a well in the centre and break 1 egg into it; add the stirred, dissolved yeast and work in by hand, adding the grated carrot. Knead for 5–10 minutes until elastic, then shape into a ball and leave to rise in the bowl in a warm place for 30–40 minutes, covered with a cloth that has been rinsed in hot water and wrung out.

When this dough has more than doubled in bulk, pour the melted butter over it, sift in the remaining flour and add the remaining eggs, the sugar, almonds, grated lemon rind and 2–3 drops of vanilla. Work the dough very thoroughly. Knead for 5–10 minutes, shape into a large ball, cover the bowl with a warm, damp cloth and leave to rise once more in a warm place in the bowl, covered with a warm, damp cloth, for 1 hour.

Have the oven preheated to 350°F (180°C) mark 4. Grease a ring mould with an outer diameter of 9½–10 in (24.5 cm) with butter; shape the dough into a fat ring that will fit into it, distribute it evenly inside the mould and smooth the surface level. Bake for 40 minutes or until the loaf is a deep russet colour on top and a skewer inserted deep into it comes out clean and dry. Take out of the oven and leave to stand for 5 minutes before unmoulding.

———•———

ONION SALAD WITH COLD MEAT

1 large or 2 medium-sized mild red onions
¾ lb (320 g) cold cooked ham or pork, cut into strips
2 tbsp finely chopped parsley
olive oil
wine vinegar or cider vinegar
salt and pepper
Serve with:
carrot, lettuce and endive salad
hard-boiled eggs

Pressed, cooked meats are best for this dish; buy them thinly sliced and cut into strips as neatly as possible.

Trim and peel the onion; cut lengthwise into quarters and then slice each quarter thinly lengthwise again into matchstick strips. Mix with the meat in a bowl, sprinkle with 2 tbsp good-quality vinegar and season with salt and freshly ground pepper. Stir and leave to stand for 5 minutes before sprinkling with approx. 3 tbsp olive oil and the parsley. Serve at room temperature with a side salad of carrots, lettuce and endive, and hard-boiled eggs.

———•———

ONIONS WITH SHARP DRESSING

1 lb (500 g) very small onions
olive oil
wine vinegar
salt and pepper

Serve with:

vegetable omelettes or cold meats

Peel the onions and remove the outermost fleshy layer. Make a cross cut 1.8 in (3 mm) deep in the base of each, as you would when preparing sprouts, and place in a large bowl of cold water. Bring a large saucepan of salted water to the boil and cook the onions until tender. Drain and place in a serving dish and sprinkle with a little vinegar, salt and freshly ground pepper. Leave to stand for 10 minutes, then sprinkle with a little olive oil. Serve warm or cold.

—•—

CARROT SALAD TURKISH STYLE
Yogurtlu Havuç Salatasi

1¼ lb (600 g) carrots

For the dressing:

3½ fl oz (100 ml) natural yoghurt

1 clove garlic, peeled and finely grated

6 tbsp olive oil

2 tbsp finely chopped parsley

salt and pepper

Peel the carrots and grate them finely. Transfer to a serving dish.

Spoon the yoghurt into a bowl, stir in the garlic and a generous pinch of salt. Beat the yoghurt continuously with a small balloon whisk or hand-held electric beater as you gradually add the olive oil a little at a time, so that it forms an emulsion. Stir in the parsley and pour this dressing over the carrot, adding a little more salt if wished, and freshly ground pepper to taste.

162

CRUDITÉS WITH SESAME DIP

6 carrots

4 green or white celery stalks

4 courgettes

4 spring onions

For the sesame dip:

7 oz (200 g) white sesame seeds

6 tbsp chicken stock (see page 37) or vegetable stock (see page 38)

4 tbsp white wine vinegar

1½ tbsp caster sugar

2 tbsp mirin or sweet white vermouth or sweet sherry

3½ tbsp Japanese light soy sauce

small pinch ajinomoto or monosodium glutamate (optional)

salt

Begin by making the sauce: spread the sesame seeds out in a large non-stick frying pan and toast over a moderate heat, stirring continuously to ensure that they cook evenly and do not burn. When they begin to jump about in the pan, remove from the heat. Put them in the food processor and reduce to a smooth, creamy paste; transfer this to a bowl and stir in the stock, vinegar, wine, soy sauce, a pinch of salt, the sugar and, if used, the *ajinomoto* or monosodium glutamate. Taste and add a little more salt if wished or a little more vinegar if you prefer a sharper taste. Keep at room temperature until ready to serve.

Clean and prepare the vegetables; cut lengthwise into fairly wide sections and then cut into large, thick matchstick shapes about 4 in (10 cm) long. Arrange the vegetables in alternating bundles on a large, round plate and place the bowl containing the sesame dip in the centre. If you need to prepare the vegetables a few hours in advance, keep them crisp in a bowl of iced water to which the juice of half a lemon has been added; take them out of the refrigerator and drain well shortly before serving.

Carrot, Celery and Fennel Salad with Prawns

3 carrots

3 celery stalks

2 bulbs fennel

12 oz (320 g) peeled large prawns

olive oil

salt and pepper

If you have bought raw prawns, steam them for 3 minutes and peel. Place the prawns in a bowl over barely simmering water, cover and leave to keep hot while you trim, peel, rinse and dry the vegetables. Use a mandoline slicer or similar utensil to slice the carrots into very thin rounds and the celery into very thin sections. Cut the fennel hearts lengthwise in half and then slice very thinly lengthwise.

Arrange the vegetables in layers on individual plates, starting with a fennel layer, followed by celery and topped with carrot. Cut the crustaceans lengthwise almost in half, open them out like a butterfly and arrange on the vegetable bed. Sprinkle with a very little salt, some freshly ground pepper and a little olive oil. Serve at once.

——•——

Daikon Root Salad with Chilli and Salmon Roe
Ikura to daikon no oroshi ae

½ daikon root

1–2 dried red chilli peppers

1 small jar salmon roe or red lumpfish roe

½ lemon

Garnish with:

sprigs of parsley

radish flowers (see page 28)

Prepare the daikon root: peel and cut it into small pieces and place in the food processor with the 1 or 2 dried chillies (remove the stalk and seeds). Process until you have a smooth, very pale pink creamy mixture, place this in the centre of a clean cloth, gather the edges together and twist round tightly to squeeze out as much moisture as possible. Heap the radish mixture on to 4 plates and garnish with parsley and radishes cut into tulip shapes. Top each serving with a spoonful of the salmon or lumpfish roe, sprinkle with a little lemon juice and serve immediately.

This delicate and unusual starter should be prepared just before it is to be eaten as the daikon mixture (*momiji oroshi*) deteriorates rapidly once grated. This is an ideal beginning to a formal Japanese meal.

——•——

Radish and Cress Salad in Cream Dressing

1 medium-sized bunch radishes

1 bunch young watercress or bitter cress or 3 boxes mustard and cress

2 small lettuce hearts

For the cream dressing:

2 tbsp white wine vinegar

pinch mustard powder

4 tbsp sunflower oil

3 tbsp double cream

pinch cayenne pepper

pinch salt

Trim and prepare the salad vegetables, rinse, drain and pat dry with kitchen paper. Tear the larger lettuce leaves into smaller pieces and arrange on individual plates with the smaller, inner leaves. Remove and discard the larger stalks from the watercress. Slide the radishes into thin rounds and spread on

163

top of the lettuce; top with your chosen type of cress.

Make the dressing: mix all the ingredients except the oil in a bowl and gradually add the oil while beating continuously to form a smooth emulsion. Sprinkle over the individual salads just before serving.

———•———

CELERIAC SALAD

1 ¾ lb (800 g) celeriac
½ lemon
approx. 10 fl oz (300 ml) mayonnaise (see page 35)
3½ fl oz (100 ml) whipping cream
salt and pepper
Serve with:
cured meats and sausages

Peel the celeriac, rinse well and dry. Cut into quarters to make it easier to slice very thinly on a mandoline cutter; reassemble each quarter's slices on top of one another and slice downwards into julienne strips (see page 24). Transfer the celeriac strips into a salad bowl, season lightly with salt and freshly ground pepper and sprinkle with the lemon juice. Stir in the mayonnaise, followed by the stiffly beaten cream. Add a little more salt and pepper if wished. This is an excellent vegetable winter salad, particularly good with a mixed platter of sliced cured meats and salami, and with cold roast meat or poultry.

———•———

RUSSIAN BEETROOT PATÉ

1¼ lb (600 g) raw beetroot
2 oz (60 g) butter, softened
1 lemon
2 tbsp caster sugar

2 small cloves garlic, peeled and finely grated
2 tbsp finely chopped parsley
salt and black pepper
Serve with:
triangles of buttered toast
Russian Carrot and Green Apple Salad (see page 167)

Scrub the beetroot under running cold water with a soft brush to remove all traces of soil; place them in a pressure cooker with a little water and cook for 20 minutes, or boil for about 2 hours, until tender, in a large pan of lightly salted water. Leave to cool before peeling. Chop very finely. Mix the beetroot in a bowl with the butter, softened at room temperature, the juice and finely grated rind of the lemon, a generous pinch of salt and the sugar. Transfer this mixture to a non-stick saucepan and cook over a moderate heat for about 7 minutes, stirring continuously, when it should have become thick. Set aside. When cold, stir in the garlic and parsley, adding freshly ground pepper and more salt if needed. Press down firmly into 4 small terrines or ramekin dishes and chill in the refrigerator for a few hours or overnight. Serve as an appetizer with toast and Russian Carrot and Green Apple Salad (see page 167).

———•———

BEETROOT AND CARDOON SALAD

2 medium-sized cooked beetroot
1 ¾ lb (800 g) cardoon hearts
juice of 1 lemon
citronette dressing (see page 35)
2 tbsp finely chopped parsley
For the *blanc de cuisine* or vegetable cooking stock:
1 oz (25 g) plain flour
juice of ½ lemon

1½ tbsp butter

2 tbsp oil

salt

Make the blanc de cuisine, which will prevent the cardoons discolouring as they boil: place the flour in a large mixing bowl and gradually stir in 2 fl oz (50 ml) cold water, blending very thoroughly with a wooden spoon until no lumps remain. Gradually stir in a further 1¾ pints (1 litre) cold water. Pour this mixture through a fine sieve into a large, deep saucepan; add 2 tsp salt, the lemon juice, the butter and the oil. Bring to the boil, stirring continuously to prevent the flour sinking to the bottom and catching; once the liquid has reached a full boil, reduce the heat and keep at a gentle boil without stirring. Trim and wash the cardoons, run the potato peeler down the outer surfaces to eliminate any strings and cut each stalk into 2½-in (6–7-cm) lengths; as each section is prepared, place it into a large bowl full of cold water acidulated with the lemon juice to prevent discoloration. When they are all ready, drain and add to the cooking liquid. Cover and boil gently for 1½–2 hours, until tender. Drain and leave to cool.

Peel the cooked beetroot and slice them. Arrange the beetroot slices and the cardoon pieces in concentric circles on a serving platter, radiating out from the centre. Sprinkle with the citronette dressing and the chopped parsley and serve.

———•———

Baked Sweet Potatoes with Cream Dressing

4 sweet potatoes, preferably red-skinned

5 fl oz (150 ml) cream dressing (see page 163)

1 large spring onion, finely chopped

½ green chilli pepper, seeds removed, finely chopped

1 tbsp coriander leaves, finely chopped

salt

Serve with:

1 lime, quartered

Have the oven preheated to 400°F (200°C) mark 6. Scrub the sweet potatoes under running cold water, dry them and wrap each separately in foil. Bake for 45 minutes or until tender. Meanwhile, prepare the cream dressing (see recipe for Radish and Cress Salad in Cream Dressing on page 163), stir in the chopped spring onion, the chilli pepper and the coriander leaves. Season to taste.

Keep the cream dressing in the refrigerator until just before serving the sweet potatoes with the foil folded back sufficiently to enable you to make a deep crosswise incision in the top of each sweet potato, in which to place 1–1½ tbsp of the cream; hand round the rest of the cream separately. Garnish each serving with a wedge of lime to be squeezed over the potato.

———•———

Turkish Potato Salad
Patates Salatasi

6 medium-sized waxy potatoes

For the wine dressing:

4½ fl oz (140 ml) olive oil

6 tbsp dry white wine

juice of 1½ lemons

2–3 tbsp finely chopped parsley

salt and pepper

Serve with:

Greek Aubergine Dip (see page 110)

Courgette Fritters (see page 143)

Scrub the potatoes under running cold water to eliminate any soil; steam them until tender, leave to cool then peel and cut lengthwise in half; cut each half into thin slices. Transfer to a deep serving dish or salad bowl. Make the dressing: dissolve scant ¾ tsp salt in the lemon juice in a bowl. Add the olive oil very gradually, beating continuously with a balloon whisk or hand-held electric beater. Keep beating as you add the wine vinegar a little at a time. Season generously with freshly ground pepper and mix carefully, coating all the potato slices. Sprinkle with the parsley and serve with a selection of Greek and Turkish appetizers such as Aubergine Dip and Courgette Fritters.

----- • -----

BAKED POTATOES WITH CAVIAR AND SOURED CREAM

4 medium-sized baking potatoes

4 tbsp caviar or black lumpfish roe (Danish caviar)

7 fl oz (200 ml) soured cream

1–1½ tbsp chopped fresh dill

salt

Garnish with:

4 lemon wedges

sprigs of fresh dill

Preheat the oven to 400°F (200°C) mark 6. Scrub the potatoes completely clean under running cold water, dry them and wrap each one in foil. Bake in the oven for 50 minutes or until tender; pull aside the foil just enough to make a deep crosswise incision right across the top of each potato, press the ends of each potato towards the middle to make the cooked potato emerge from the cut skin and top each with a pinch of salt, chopped dill, 2 tbsp soured cream and the caviar or lumpfish roe. Serve at once, garnished with the lemon wedges and sprigs of dill.

----- • -----

HUNGARIAN POTATO SALAD

4 large waxy potatoes

4 hard-boiled eggs

2 spring onions

4 fl oz (120 ml) mayonnaise (see page 35)

soured cream

salt and pepper

Serve with:

green salad

Peel the potatoes, cut them into small pieces and place in a saucepan with 18 fl oz (500 ml) water and ½ tsp salt. Bring to the boil, cover tightly and cook over a low to moderate heat for about 15–20 minutes or until the potatoes are tender and have absorbed all the water. Chop the hard-boiled eggs coarsely; trim the spring onions, removing the outer layer, and slice the bulb and leaves into rings. Use a fork to partially break up the hot, cooked potato pieces. Leave to cool at room temperature.

Make the mayonnaise, stir in 3 tbsp soured cream or smetana and mix with the potatoes, the spring onions and chopped eggs. Season with salt and freshly ground pepper to taste. Serve at room temperature, with green salad, as an appetizer or as part of a cold lunch.

----- • -----

JERUSALEM ARTICHOKE SALAD

1¼ lb (600 g) Jerusalem artichokes

2 lemons

extra-virgin olive oil

salt and pepper

Garnish with:

½ tbsp finely chopped parsley

1 lemon, thinly sliced into rounds

Have a bowl full of cold water acidulated with the juice of 1 lemon standing ready; peel the Jerusalem artichokes and immediately drop into the water to prevent discoloration. This stage of the preparation can be done in advance.

Shortly before serving, slice the drained and dried raw artichokes very thinly on a mandoline cutter or in your food processor. Spread out the slices, overlapping one another, on individual plates; season with salt, freshly ground pepper and sprinkle with a little olive oil and lemon juice. Cut the lemon slices in half and arrange some on each plate, in a fan shape. Serve immediately.

—•—

RUSSIAN CARROT AND GREEN APPLE SALAD

6 carrots
3 crisp green apples
juice of 2 lemons
1 egg
7 fl oz (200 ml) olive oil
½ tsp Dijon mustard
Worcester sauce
salt and pepper
Garnish with:
2 tbsp soured cream
4 sprigs fresh chervil

Peel the carrots and apples and grate them coarsely; mix together in a bowl with a pinch of salt and half the lemon juice. Make the mayonnaise as follows: break the egg into the blender (this method uses both yolk and white), add the mustard, the juice of the remaining half lemon, 4–6 drops Worcester sauce and a little salt and pepper. Blend at high speed while gradually pouring the olive oil in a thin stream into the blender receptacle through the hole in its lid; the mayonnaise will become thick, pale and

smooth. Add a little more salt, pepper and lemon juice to taste and mix with the grated vegetables; chill briefly before serving in individual bowls, topping each serving with a little soured cream and a small sprig of fresh chervil.

—•—

GRATED CARROT SALAD WITH INDIAN MUSTARD DRESSING

10 medium-sized carrots
2 tbsp sultanas
2 tbsp fresh lime juice
4 tbsp sunflower oil
1½ tbsp whole black mustard seeds
salt

Soak the sultanas in warm water for 30 minutes, then drain well. Peel the carrots and grate coarsely. Dissolve scant ¾ tsp salt in the lime juice in a large mixing bowl. Heat the sunflower oil in a small saucepan and when very hot, add the mustard seeds; fry for just a few seconds, pour all over the grated carrot. Stir well and serve.

—•—

CARROT AND ONION SALAD WITH LIME AND GINGER DRESSING

1¼ lb (600 g) carrots
1 medium-sized mild red onion
½ green chilli pepper, seeds removed, finely chopped
2 tbsp fresh lime or lemon juice
4 tbsp sunflower oil
1 small piece fresh ginger, peeled and finely grated
salt

Peel the carrots and grate coarsely. Peel the onion and cut into thin strips. Mix the carrot and onion with the chopped chilli pepper in a large bowl. Dissolve a large pinch of salt in the lime juice in a small bowl, then gradually beat in the oil adding a little at a time to form an emulsion like a vinaigrette. Stir in the grated ginger. Pour over the carrot and onion mixture, mix briefly and leave to stand for 10–15 minutes. Serve at room temperature.

—— • ——

JAPANESE DAIKON ROOT AND CABBAGE SALAD
Namasu

1 medium-sized daikon root

¼ firm green or white cabbage

2 small tender carrots

For the *kimi zu* dressing:

1 tbsp cornflour or potato flour

4 fl oz (120 g) cold Japanese instant stock (dashinomoto)

4 egg yolks

4 tbsp Japanese rice vinegar or diluted cider vinegar (see method)

1 tbsp caster sugar

salt

Cut the vegetables into strips. Make the dressing: heat some water in the bottom of a double boiler or saucepan. Place the cornflour or potato flour in a bowl and gradually stir in the cold *dashinomoto* or stock, adding a little at a time to prevent lumps forming. Beat the egg yolks continuously in the top of the double boiler or in a heatproof bowl over the simmering water, using a balloon whisk or hand-held electric beater as you add the cornflour mixture a little at a time. Add the sugar, ½ tsp salt and continue beating as you gradually add the vinegar. The mixture will eventually start to thicken; as soon as it does, remove from the heat and leave to cool. Cider vinegar is best diluted when substituted for Japa-

nese rice vinegar: mix 3 tbsp of it with 1 tbsp cold water for this recipe and add a pinch of extra sugar. When the vegetables have crisped in the refrigerator, drain well and dry in a salad spinner or with kitchen paper; mix well with the dressing and serve.

—— • ——

JAPANESE MIXED SALAD WITH SAMBAI ZU DRESSING

½ daikon root

2 young carrots

2 crisp lettuce hearts

2 green celery stalks

2 thick slices lean cooked ham (approx. 8 oz (225 g) total weight)

For the *sambai zu* dressing:

2 tbsp Japanese light soy sauce

2 tbsp Japanese rice vinegar (or cider vinegar, see method)

6 tbsp sunflower oil

salt and freshly ground pepper

Trim and peel the vegetables, wash and drain well. Tear the larger lettuce leaves into small pieces. Use a mandoline cutter or food processor to slice the daikon root and carrot into wafer-thin rounds and the celery into very thin pieces cutting across the stalk. Mix all these in a large bowl. Cut the ham into strips. Make the dressing: mix 2 pinches of salt, a little freshly ground pepper, the soy sauce and the vinegar in a bowl. If using cider vinegar, use 1½ tbsp and add ½ tbsp of cold water and a generous pinch of caster sugar. Keep beating with a balloon whisk as you gradually add the oil to form an emulsion. Add this dressing to the salad and mix well; transfer to individual plates and place four ham strips on top of each serving.

—— • ——

Potato Salad with Spring Onions and Chives

3 large waxy potatoes

2 spring onions

10 fl oz (300 ml) single or soured cream

2 tbsp coarsely chopped chives

salt and freshly ground black pepper

Peel the potatoes, cut them into small pieces and place in a saucepan with 18 fl oz (500 ml) water and ½ tsp salt. Bring to the boil, cover and simmer over a very low heat for 15–20 minutes or until the potatoes are tender and have absorbed all the water. Use a fork to break up the potatoes into very small pieces. Leave to cool but do not chill.

Pour the cream over the potatoes. Remove the outer layer of the spring onions and slice the bulbs into thin rings. Add to the potatoes and cream together with plenty of freshly ground black pepper and a little more salt to taste. Mix well. Transfer to a serving bowl, sprinkle with the chives and serve.

———— • ————

Chilled Leek and Potato Soup
Vichyssoise

1 lb (500 g) leeks (net trimmed weight), white part only

14 oz (400 g) floury potatoes (net peeled weight)

2 pints (1.2 litres) chicken stock (see page 37)

9 fl oz (250 ml) single cream

3 tbsp chopped chives

salt and freshly ground white pepper

Cut the potatoes into small pieces and the white part of the leeks into very thin rings; place both in a saucepan with the stock, bring slowly to the boil, cover and simmer for about 50 minutes or until very tender. Use a hand-held electric beater to beat until smooth or put through the vegetable mill. Stir in the cream and add salt and pepper to taste.

When cool, chill in the refrigerator and serve very cold, garnishing the individual servings with chopped chives.

———— • ————

Cream of Celeriac Soup

1 lb (500 g) celeriac

1¾ pints (1 litre) chicken stock (see page 37)

1 oz (30 g) butter

1 oz (30 g) rice flour

3½ fl oz (100 ml) single or double cream

2 egg yolks

1 lemon

1–1½ tbsp finely chopped parsley

salt and freshly ground white pepper

Serve with:

croutons

Heat the stock. Wash, dry, peel and dice the celeriac; place in a saucepan and add about half the boiling hot stock, sufficient to cover it. Cover and boil gently until the celeriac is very tender; put through a vegetable mill to purée. Heat the butter over a low heat in a very large, heavy-bottomed saucepan; stir in the rice flour and cook for 1–2 minutes or until it begins to turn a pale golden brown. Draw aside from the heat and beat in the remaining hot stock. Return to the heat and bring very slowly to the boil while stirring continuously. Simmer for a further 5 minutes before stirring in the celeriac purée. Reduce the heat to very low. Beat the egg yolks and cream them together in a small bowl or jug until blended; beat this thickening liaison into the soup, adding a little at a time. Continue to beat for a few minutes, until the soup is very hot, but do not allow to boil or it will curdle. Draw aside from the heat, stir for 1–2 minutes and season to taste with salt, pepper and a little lemon juice. Sprinkle with parsley and serve with the croutons, fried in butter until golden brown.

———•———

CREAM OF PARSNIP SOUP WITH BACON AND GARLIC CROUTONS

1 lb 2 oz (500 g) parsnips, trimmed (peeled net weight)

2 rashers smoked streaky bacon, ¾ in (2 cm) thick

2 oz (50 g) unsalted butter

2 shallots, peeled and finely chopped

3½ fl oz (100 ml) dry or medium-dry white wine

1½ pints (800 ml) chicken stock (see page 37)

3 egg yolks

9 fl oz (250 ml) single cream

nutmeg

salt and freshly ground white pepper

For the garlic croutons:

5 small slices white bread

2 tbsp olive oil

2 cloves garlic

salt

Make the garlic croutons. Trim off and discard the crusts from the bread slices and cut into small squares or dice. Heat the olive oil and the peeled, whole garlic cloves in a very wide, non-stick frying pan, add the diced bread and fry over a moderate heat until crisp and golden brown all over, turning them as required. Sprinkle them with a little salt and set aside.

Use streaky bacon with a good proportion of lean to fat, remove the rind and excess solid fat and cut the bacon into very small dice. Fry over a moderate heat in the non-stick pan without any additional oil, turning now and then, until the bacon is crisp and lightly browned; take the bacon out of the pan and blot up excess fat from the dice with kitchen paper. Set aside.

Dice the peeled parsnips. Heat half the butter in a large, heavy-bottomed saucepan and sweat the chopped shallot and the diced parsnip in it, stirring frequently, for 10 minutes. Increase the heat from low to moderate, add the wine and cook, uncovered, until it has completely evaporated. Add the stock, a small pinch of salt and bring to the boil. Simmer uncovered over a moderate heat for about 15 minutes or until the parsnips are very tender. Reduce the heat to low. Beat the egg yolks and cream together briefly and gradually add this mixture to the soup while beating continuously. Continue beating until the soup is very hot but do not allow it to boil or it will curdle. Remove from the heat and beat with a hand-held electric beater until smooth; beat in the remaining, chilled butter a small piece at a time. Add a little more salt if necessary (not too much, as the bacon will add more), freshly grated white pepper and a little grated nutmeg. Ladle into hot individual soup bowls, sprinkle the garlic croutons and crispy bacon over the surface and serve at once.

———•———

PHILADELPHIA PEPPER-POT

3 medium-sized potatoes, diced

½ medium-sized onion, peeled and finely chopped

1 green celery stalk, finely chopped

1 large green pepper, finely chopped

1 oz (30 g) butter

2 tbsp olive oil

3 tbsp plain flour

generous 2 pints (1.2 litres) chicken stock (see page 37)

7 oz (200 g) cooked honeycomb tripe, diced

5 fl oz (150 ml) double cream or soured cream

salt and freshly ground black pepper

Serve with:

hot crusty bread

Heat the stock. Prepare the vegetables. Fry the onion, celery and pepper gently in the butter and oil for 15 minutes, stirring frequently. Add the flour and cook, stirring, over a low heat for 2 minutes. Continue stirring as you gradually add the hot stock, followed by the tripe, potatoes and plenty of coarsely ground pepper. When the liquid comes to the boil, cover and simmer gently for about 1 hour. Add salt to taste and remove from the heat. Ladle into individual soup bowls, spoon 2 tbsp soured cream into the centre of each serving and finish with a little more freshly ground pepper.

———•———

RUSSIAN BEETROOT AND CABBAGE SOUP
Borscht

12 oz (350 g) raw beetroot

7 oz (200 g) white cabbage

1 carrot

1 leek

1 medium-sized potato

1 green celery stalk

1½ oz (40 g) butter

12 oz (350 g) boiling beef (e.g. top rump or leg)

1 thick slice smoked streaky bacon

¼ lb (100 g) fresh or drained tinned tomatoes

1 bay leaf

2 tbsp chopped fresh dill

7 fl oz (200 ml) soured cream or smetana

1 tbsp red wine vinegar

salt and black peppercorns

Trim, wash and dry all the vegetables. Peel the beetroot, reserving 4 oz (100 g) in weight and cut all the rest, and the cabbage, carrot, leek and celery into strips. Dice the peeled potatoes.

Heat the butter in a stockpot or a very large, deep, enamelled cast iron casserole dish and fry the vegetables in it gently for about 15 minutes, stirring frequently. Add 2 pints (1.2 litres) water, the chopped, seeded tomatoes and the beef. Add the bay leaf, bacon, a few black peppercorns and ¾ tsp coarse sea salt. Bring to the boil, skimming off any scum that rises to the surface, then turn down the heat to very low and simmer very gently for 3 hours, or until the beef is extremely tender.

Cut the reserved raw beetroot into small pieces and grate finely, place in a piece of muslin and twist tightly to force out all the juice, collecting this in a small bowl. Stir in a pinch of salt and the vinegar.

Take the beef and bacon out of the soup, spearing them with a carving fork and cut into strips. Return these to the soup. Add a little salt if necessary and freshly ground black pepper to taste. Draw aside from the heat, stir in the raw beetroot juice and vinegar mixture and immediately ladle into individual soup bowls. Place 1–1½ tbsp chilled soured cream in the centre of each serving and sprinkle with the chopped dill.

———•———

ICED BEETROOT SOUP

9 oz (250 g) cooked beetroot

½ medium-sized onion, coarsely chopped

2 medium-sized floury potatoes, steamed or boiled in their skins until tender

16 fl oz (450 ml) chicken stock (see page 37)

2 tbsp dry white wine or dry vermouth

1 lemon

9 fl oz (250 ml) natural yoghurt

salt and pepper

Garnish with:

5 fl oz (150 ml) whipping cream

1 tbsp chopped chives

Place the onion in a very large saucepan with 7 fl oz (200 ml) of the stock and the wine or vermouth, cover and boil gently over a low heat for 15 minutes. Remove the lid and increase the heat to moderately high to reduce the liquid for 2–3 minutes. Draw aside from the heat.

Peel the potatoes and beetroot, slice very thinly and add to the saucepan together with a further 9 fl oz (250 ml) of the stock, the lemon juice, a generous pinch of salt and some freshly ground pepper. Beat well with a hand-held electric beater until smooth and creamy or put through a vegetable mill. Stir in the yoghurt and a little more salt to taste (the soup will taste less salty when chilled).

Pour into individual soup plates, cover with cling film and place in the refrigerator until just before serving, decorating the well chilled soup with a spoonful of whipped cream and some chopped chives.

———— • ————

CREAM OF POTATO AND LEEK SOUP
Potage Parmentier

1 lb (500 g) peeled floury potatoes

1 lb (500 g) leeks (net trimmed weight)

4 tbsp double cream

3 tbsp finely chopped parsley

scant 1 oz (20 g) unsalted butter

salt and white peppercorns

Cut the potatoes into small pieces and slice the well-washed leeks into thin rings (use both green and white parts but remove and discard the outer layer and the tougher, darker green leaves). Place these vegetables in a large saucepan with scant 2½ pints

(1.4 litres) water, 1–1½ tsp salt, bring slowly to the boil and simmer, partially covered to allow the steam to escape, for 50 minutes. Use a hand-held electric beater to beat until smooth and creamy or put through a blender or vegetable mill.

The soup can be prepared in advance to this point and reheated just before serving.

Stir the cream into the very hot soup off the heat; stir in the butter, adding a small piece at a time, followed by the parsley. Taste and add a little more salt if needed, and some freshly ground white pepper.

———— • ————

CREAM OF LEEK AND COURGETTE SOUP

2¼ lb (1 kg) leeks

1½ lb (700 g) courgettes

2 chicken stock cubes

2 tbsp finely chopped parsley

7 fl oz (200 ml) single cream

salt and pepper

Prepare the leeks, removing the outer layer and trimming off the ends; use both the white and the tender green parts. Wash very thoroughly and cut into thick rings. Trim and wash the courgettes; slice into thick rings. Place the vegetables in a large saucepan with 2¾ pints (1.6 litres) water and the 2 crumbled stock cubes. Bring to the boil, cover and simmer over a low heat for 50 minutes.

Turn off the heat and beat with a hand-held electric beater until smooth, or put through a blender or vegetable mill; return to the pan and reheat briefly. Add salt and freshly ground white pepper to taste and stir in the parsley.

Ladle into soup bowls. Pour the cream in a very thin stream into each bowl to form concentric circles then draw the tip of a spoon from the centre out-wards across the circles to give an attractive spider's web effect and serve hot or chilled.

Cream of Potato Soup with Crispy Leeks

1¼ lb (600 g) floury potatoes, peeled

generous 1 pint (700 ml) chicken stock (see page 37)

9 oz (250 g) trimmed leeks, white part only (net weight)

1 oz (30 g) butter

9 fl oz (250 ml) single cream

1–1½ tbsp finely chopped parsley

generous pinch caster sugar

nutmeg

salt and pepper

Slice the potatoes very thinly and place in a large saucepan with the stock. Bring to the boil, cover and simmer gently over a low heat for 30 minutes until very tender. Put through a blender or vegetable mill; return this purée to the saucepan.

Slice the leeks into very thin rings and sweat over a very low heat in the butter until only just tender and still crisp. Season with a little salt, freshly ground pepper and a pinch each of sugar and grated nutmeg.

Stir the cream into the thin potato purée and heat to just below boiling point. Add a little salt and pepper to taste. Transfer this soup to individual bowls, spoon equal amounts of the leek into the centre of each serving, sprinkle with a little parsley and serve.

———— • ————

Onion Soup

1¾ lb (800 g) trimmed and peeled onions (net weight)

scant 2¾ pints (1½ litres) beef stock (see page 37)

12 thick slices French bread

1½ oz (40 g) butter

3 tbsp olive oil

generous pinch caster sugar

4 tbsp plain flour

3½ fl oz (100 ml) dry vermouth or dry white wine

3 tbsp brandy

1 large clove garlic, peeled

4 oz (110 g) grated Gruyère or Emmenthal cheese

salt and freshly ground pepper

Heat the stock. Preheat the oven to 325°F (170°C) mark 3 and bake the bread slices on a baking tray for 30 minutes to make them dry and crisp.

Cut the onions in half from top to bottom; slice very thinly, also from top to bottom.

Sweat the onions over a very low heat for 15 minutes in the butter and 1 tbsp of the oil in a large, heavy-bottomed saucepan with the lid on. Remove the lid, sprinkle with 1 tsp salt and a generous pinch of sugar (the latter will help them brown) and continue cooking over very low heat, taking the lid off for frequent stirring. When the onions are golden brown, add the flour and cook, stirring with a wooden spoon for about 2 minutes.

Remove the saucepan from the heat and keep beating continuously and vigorously with a balloon whisk as you gradually add the boiling hot stock, to prevent lumps forming. Add the vermouth or dry white wine and stir.

Return the saucepan to a low heat, cover, leaving a gap for the steam to escape and simmer for 40 minutes; carefully skim off any skin which may form on the surface at intervals. Add a little more salt if wished and the brandy.

When the bread slices have had 20 minutes in the oven, take them out, sprinkle with the remaining oil and return to the oven for another 10 minutes, then take them out again, cut the garlic clove in half and rub the cut surfaces firmly all over the crisp bread. Place 2 in each of 6 heated soup bowls, ladle the soup over them and serve, handing round the grated cheese separately. Serves 6.

———— • ————

ONION SOUP AU GRATIN

Ingredients as for Onion Soup on page 173 with the addition of:

1½ tbsp grated raw mild red onion
3 oz (90 g) freshly grated Parmesan cheese
2 oz (60 g) Gruyère or Emmenthal cheese
2 tbsp olive oil

Make the onion soup (see page 173). Mix the 3 oz (90 g) freshly grated Parmesan cheese listed above with the 4 oz (110 g) grated Gruyère or Emmenthal cheese listed in the recipe on page 173 and set aside. Cut the extra Gruyère or Emmenthal cheese listed above into wafer-thin flakes. Preheat the oven to 325°F (170°C) mark 3.

Pour the soup into ovenproof bowls. Sprinkle the cheese flakes and the grated raw onion evenly over the surface, place 2 crisped bread croutons (see recipe on page 173) in each bowl of soup and sprinkle them with the grated cheeses. Drizzle a little olive oil over the cheese and finish with a generous grating of black pepper. Place on the top shelf of the oven for 20 minutes then place the bowls under the grill for 5 minutes. Serves 6.

———— • ————

PROVENÇAL GARLIC SOUP
Aïgo Bouido

17–20 cloves garlic
2 cloves
1 fresh sage leaf
generous pinch fresh thyme
1 bay leaf
5 sprigs parsley
3 egg yolks
4 tbsp olive oil

3 oz (60 g) grated Parmesan cheese
3 oz (60 g) grated Gruyère or Emmenthal cheese
salt and pepper

Separate the garlic cloves, detaching them from the base; do not peel them. Blanch them for 30 seconds in a saucepan of fast boiling unsalted water; drain, refresh under running cold water and peel. Put them in a large saucepan with 2½ pints (1.4 litres) water, a large pinch of salt and some freshly ground pepper, the cloves, sage leaf, thyme, bay leaf and parsley sprigs. Bring to the boil then simmer for 30 minutes; put through the vegetable mill with a fine-gauge sieving disc fitted; return this soup to the saucepan, add a little more salt if needed, heat to boiling point, and then turn off the heat. Beat the egg yolks in a bowl until they are smooth and creamy; continue beating as you gradually add the olive oil a little at a time. Gradually beat in about 4 fl oz (120 ml) of the hot soup, then reverse the procedure and gradually beat the contents of the bowl into the saucepan containing the hot soup. Make sure the liquid in the pan is no longer near boiling point. Serve with the mixed grated cheeses.

———— • ————

SALSIFY AND MONKFISH SOUP

1½ lb (700 g) scorzonera (black salsify)
14 oz (400 g) monkfish
juice of 1 lemon
1¾ pints (1 litre) fish stock (see page 38)
3 tbsp olive oil
1 clove garlic
1 sprig thyme
salt and pepper

Serve with:

garlic croutons (see page 170)

Black salsify discolours very quickly once peeled, so have a large bowl of cold water acidulated with half the lemon juice standing ready and drop each piece into it as soon as it is peeled. Cut the salsify into pieces 1½ in (4 cm) long and boil for 20 minutes in a large saucepan of salted, acidulated water. Drain, leave to cool and then cut each piece lengthwise in half, remove the woody centre and slice the remaining, tender parts into strips. Take the monkfish flesh off the bone, removing all the skin, and cut into fairly large pieces. Sprinkle with a little salt. Heat the fish stock.

Fry the peeled whole garlic clove and the thyme gently for 2 minutes; add the fish pieces and sauté for 1 minute. Add the salsify and continue to fry gently for a further 2 minutes while stirring. Add the hot fish stock, bring slowly to a very gentle boil and allow to simmer for 30 seconds only before removing from the heat. Check that the fish is cooked (simmer for a little longer if not), add a little more salt if necessary, some freshly ground pepper and serve with the garlic croutons.

——— • ———

ICED HAWAIIAN SOUP

2 large floury potatoes

12 fl oz (350 ml) chicken stock (see page 37)

9 fl oz (250 ml) dry white wine

1 large or 2 small cucumbers

4 tender green celery stalks

1 small onion

9 fl oz (250 ml) single cream

1 lime

salt and black pepper

Garnish with:

1½–2 tbsp finely chopped coriander or parsley

Peel the potatoes, cut them into small pieces and place in a large saucepan with the stock and the wine. Bring quickly to the boil over a high heat, then reduce the heat, cover and simmer for 20 minutes. While the potatoes are cooking, peel and prepare the remaining vegetables. Chop them all finely together, then add to the potatoes and continue simmering for a further 10 minutes. Remove the saucepan from the heat and purée the soup in a blender. Add salt and, when cool, stir in the cream and season with black pepper. Serve chilled with the lime cut lengthwise into quarters for each person to squeeze into the soup; decorate with chopped coriander or parsley.

——— • ———

SPAGHETTI WITH SPRING ONION, TOMATO AND HERB SAUCE

1½–1¾ lb (750 g) large spring onions

2 cloves garlic, peeled and finely chopped

1 bouquet garni

7 tbsp olive oil

8 oz (225 g) ripe tomatoes, skinned and seeded

1–2 red chilli peppers

12 oz (320 g) spaghetti

8 basil leaves

salt

Slice the spring onions. Place in a large fireproof casserole dish with the chopped garlic, the bouquet garni and 4 tbsp of the olive oil. Sweat the spring onions over a very low heat, uncovered, for 20 minutes, stirring frequently, until soft and reduced to less than one-quarter of their original volume.

Bring a large saucepan of salted water to the boil ready to cook the spaghetti. Discard the bouquet garni. Chop the tomatoes coarsely. Remove the stalk and seeds from the chilli peppers and chop the flesh very finely. Add the tomatoes and chilli pep-

175

pers to the spring onions with a pinch of salt and cook gently for a further 5 minutes, stirring occasionally. Remove from the heat.

Cook the spaghetti until just tender, with some 'bite' left to it; reserve 2–3 fl oz (60–90 ml) of the cooking water when you drain it. Add the spaghetti to the spring onion mixture, together with the reserved hot water, and stir briefly over a low heat to coat and flavour the spaghetti. Draw aside from the heat, add the remaining olive oil and the torn or shredded basil leaves. Serve at once.

———•———

PASTA AND POTATOES

4 medium-sized waxy potatoes, diced
7 oz (200 g) large pasta shapes (e.g. shells or large ribbed macaroni)
1 large onion, peeled and finely chopped
1 green chilli pepper, seeds removed, finely chopped (optional)
2 large ripe tomatoes
5 fl oz (150 ml) hot chicken stock (see page 37)
4 fl oz (120 ml) olive oil
oregano
salt and pepper
Garnish with:
sprigs of basil
4 tbsp additional diced ripe fresh tomato flesh

Bring a large pan of salted water to the boil ready to cook the pasta. Heat 6 tbsp of the oil in a wide, deep fireproof casserole dish and sweat the onion and chilli pepper, if used, over a low heat for 10 minutes, stirring frequently. Add the diced potatoes and continue to cook over a low heat while stirring for 5 minutes. Chop the tomato flesh coarsely and stir into the saucepan with a generous pinch each of oregano, salt and freshly ground pepper; stir and turn

all the ingredients while cooking for 2–3 minutes. Add the stock and stir again. Cook and simmer for 5 minutes or until the potato is just tender. Add more salt to taste. Turn off the heat.

When the pasta is tender but has still plenty of 'bite' left in it, drain, reserving about 8 fl oz (220 ml) of the cooking water. Add the pasta to the contents of the saucepan and stir over a low heat for about 5 minutes. This should complete the cooking of the pasta. Add some of the reserved cooking water if the liquid in the pan reduces appreciably, as the finished dish should be very moist. Turn off the heat and leave to stand, uncovered, for 10 minutes, to allow the flavours to blend, then transfer to heated bowls. Drizzle about ½ tbsp olive oil over each serving, season with a little freshly ground pepper and garnish with sprigs of fresh basil and some diced tomato flesh.

———•———

PISSALADIÈRE

3¼–3½ lb (1½ kg) onions
1 lb (500 g) pizza dough (see pages 70–71)
5 fl oz (150 ml) olive oil
20 small black olives
generous pinch caster sugar
salt and pepper

Make the pizza dough and while it is rising, peel and slice the onions into slivers.

Heat the oil in a very wide frying pan and fry the onions gently over a low heat for 10 minutes, stirring continuously. Add a pinch of sugar, season with salt and freshly ground pepper, cover and sweat over a very low heat for a further 15 minutes, stirring occasionally. Remove the lid, turn up the heat a little and fry gently, stirring all the time: the onions will turn a pale golden brown and shrink. Draw aside from the heat and add more salt and pepper if necessary.

Preheat the oven to 475°F (240°C) mark 9. Lightly oil a shallow, rectangular baking tray measuring approx. 14 × 16 in (35 × 40 cm), or a large quiche dish. When the pizza dough has risen sufficiently, knock it down by hand and then roll out with a lightly floured rolling pin to a rectangle large enough to line the base and shallow sides of the baking tray, or a circle to fit the quiche dish; press the dough into the corners or angles of the tray or dish with your knuckles. Sprinkle the surface with a little salt and olive oil. Spread the cooked onions evenly over the surface and arrange the olives on top. Bake for 15 minutes or until golden brown.

———•———

POTATO DUMPLINGS WITH BASIL SAUCE
Gnocchi al pesto

2¼ lb (1 kg) floury potatoes
7 oz (200 g) plain flour
1 egg
5 tbsp grated Parmesan cheese
unsalted butter
nutmeg
salt and pepper
For the *pesto* (Genoese basil sauce):
1 large bunch fresh basil
1½ oz (40 g) freshly grated Pecorino cheese
1½ oz (40 g) freshly grated Parmesan cheese
3 tbsp pine nuts
1 small clove garlic, peeled
4 tbsp extra-virgin olive oil
sea salt

Steam or boil the potatoes. While they are cooking, make the basil sauce: take all the leaves off the stalks, rinse the leaves and blot dry with kitchen paper. Place in the food processor with the cheeses, garlic, pine nuts and a pinch of sea salt. Process to a thick, smooth paste.

Peel the potatoes while they are still boiling hot and mash them with a potato ricer or by pushing them through a sieve into a large mixing bowl, working quickly so that they are still hot and easy to push through. Use your hands to mix in the sifted flour a little at a time, adding a pinch of salt and pepper, a pinch of grated nutmeg, the finely grated Parmesan cheese and the egg. The resulting potato dough should be firm, smooth and easy to shape.

Break off about a quarter of the dough; using the palms of your hands roll it on a lightly floured pastry board or working surface into a long cylinder or sausage about ¾ in (2 cm) in diameter. Cut into 1-in (2½-cm) lengths. Repeat this process with the remaining dough. Press these pieces one by one against the prongs of a fork, forming a hollow on one side of each dumpling; the other side will be rounded and patterned with the prongs. As you shape them, place on a clean cloth or board dusted with sifted flour. Sprinkle lightly with sifted flour.

Shortly before you plan to serve the gnocchi, bring plenty of salted water to an even, gentle boil in a very wide, deep saucepan. Add the gnocchi to the boiling water in batches. They will soon bob up to the surface: this means they are done; remove with a slotted spoon and place in a colander on a plate beside the saucepan. As soon as the last batch has cooked, transfer the gnocchi to very hot individual plates, top with a little unsalted butter and the basil sauce.

———•———

GENOESE SEAFOOD AND VEGETABLE SALAD
Cappon magro

For the vegetable assortment:
1 medium-sized cooked beetroot
8 oz (225 g) scorzonera

177

1 small cauliflower
4 very young, tender globe artichoke hearts or 8 artichoke bottoms
8 oz (225–250 g) French beans
5 medium-sized young carrots
1 bunch green celery, inner stalks only
2 medium-sized waxy potatoes
3–4 lemons
olive oil
salt and pepper
For the bread and seafood assortment:
1 large loaf white bread
2–3 cloves garlic
12 unpeeled, cooked large Mediterranean prawns
1 cooked lobster weighing approx. 2½ lb (1 kg)
1 cooked sea bass
olive oil
lemon juice
salt and pepper
For the sauce:
2 thick slices white bread
1 small bunch parsley
1 clove garlic
2½ oz (75 g) pine nuts
1 oz (25 g) capers
4 tinned anchovy fillets, drained
2 hard-boiled eggs, yolks only
6 large fleshy black olives, stoned
3½ fl oz (100 ml) wine vinegar
9 fl oz (250 ml) olive oil
salt and pepper

You will need extra olive oil and wine vinegar in addition to the basic quantities listed above. Trim, prepare and wash all the raw vegetables; cook each type separately. Divide the cauliflower into florets and boil in salted water. Remove the outer leaves from the artichokes, cut the hearts lengthwise into quarters then prepare and cook as described on page 105. If you cannot obtain baby artichokes, strip down to the bottoms and remove their chokes. Boil the French beans. The carrots, celery stalks and the potatoes can, however, be boiled together until tender. Drain the vegetables. Cut the celery and scorzonera into small pieces; slice the potatoes thickly. Slice the carrots then peel and slice the hot, cooked beetroot. Place each vegetable in a separate serving dish, drizzling a mixture of lemon juice, olive oil and salt and pepper over them. Set aside until required. Cut two thick, lengthwise slices from a 2–3-day-old large white loaf, and heat gently in the oven until very dry, hard and crisp. Cut a large garlic clove in half and rub the exposed surfaces against the surface of the bread slices; place these in the bottom of a deep serving dish; moisten with wine vinegar diluted with a little cold water and a little salt. Leave to soften for 2–3 hours.

Peel the prawns; take the lobster out of its shell and cut into round slices about ¼ in (½ cm) thick. Poach the sea bass in fish stock; leave to cool in the cooking liquid, drain and remove the bones. Slice as neatly as possible. Drain the oil from the tinned anchovy fillets. Keeping the different types of seafood separate, dress each with a mixture of olive oil, some lemon juice, a little salt and freshly ground white pepper to taste.

Make the sauce: cut the crusts off the bread, moisten thoroughly with vinegar and squeeze out excess moisture. Chop all the other ingredients listed for the sauce together very finely, including the moist bread; push the resulting mixture through a fine sieve into a bowl. Stir in 3½ fl oz (100 ml) vinegar and then gradually beat in the 9 fl oz (250 ml) olive oil, adding a little at a time. Season with salt and pepper. A couple of hours before assembling the salad, drizzle a little olive oil over the 2 large rectangles of bread; moisten with a layer of the sauce. Leave to stand and soften. When you are ready to assemble the salad, shortly before serving, continue layering, using one type of vegetable after another and alternating vegetables with layers of sea bass, anchovies and lobster slices. Each layer should be covered with a thin layer of the sauce. Keep the prawns until last, to decorate the surface.

Jerusalem Artichoke Vol-au-Vents Russian Style

1 lb (450 g) Jerusalem artichokes

2 shallots, finely chopped

1 oz (30 g) butter

1 tbsp olive oil

approx. 5 fl oz (150 ml) vegetable stock (see page 38)

4 vol-au-vent puff pastry cases, 6 in (15 cm) in diameter

½–1 red chilli pepper, finely chopped

6 tbsp concentrated tomato purée

10–11 fl oz (300 ml) double cream

juice of 2½ lemons

salt

Squeeze the juice of 1 lemon into a large bowl of cold water. Peel the Jerusalem artichokes, and drop them immediately into the acidulated water to prevent discoloration. When they are all prepared, drain, dry and cut into ¼-in (½-cm) dice.

Sweat the chopped shallot in ½ oz (15 g) of the butter and 1 tbsp oil, add the diced artichokes, sprinkle with a pinch of salt and cook over a low heat, stirring, for 2–3 minutes. Add the stock, bring slowly to the boil, cover and simmer gently for about 10 minutes or until they are tender but not at all mushy. Set aside.

Place the vol-au-vent cases in the oven, preheated to 325°F (170°C) mark 3, for 15 minutes. While they are heating make the sauce. Melt the remaining butter in a small saucepan and fry the chilli pepper gently for 1 minute. Add the tomato purée and continue stirring over the low heat for about 2 minutes. Stir in 7 fl oz (200 ml) boiling hot water and simmer, uncovered, over a slightly higher heat for 2–3 minutes. Stir in the cream and a pinch of salt; simmer, stirring continuously, for a further 2–3 minutes. Add this rather thin sauce to the artichokes and place them over a low heat, stirring carefully without breaking them up. When they are very hot, turn off the heat, add the remaining lemon juice, a little

more salt if needed and stir once more. Fill the pastry cases with this mixture, put their lids on top and serve at once on very hot plates.

———•———

Spanish Omelette
Tortilla de papas

4 medium-sized Spanish onions

2 medium-sized cold boiled potatoes

scant 1 oz (20 g) butter

3 tbsp olive oil

½–1 green chilli pepper or ½ red or green pepper

2 rashers smoked streaky bacon

8 eggs

4 tbsp cream

salt and pepper

Serve with:

green salad or tomato salad

Trim and peel the onions, cut from top to bottom in quarters and then slice very thinly in the same direction. Heat the butter and oil in a very wide, fairly deep frying pan (preferably non-stick) and fry the onion gently with the chilli pepper or the red peppers for 15 minutes over a low to moderate heat, stirring frequently. When they are tender and lightly browned and have shrunk considerably, add the diced potatoes, a pinch of salt, some freshly ground pepper and stir over a low heat for 2–3 minutes. Remove the rind from the bacon and cut it into small dice. This stage of the omelette preparation can be completed several hours in advance.

When you are ready to serve the omelette, fry the diced bacon in a non-stick frying pan over a high heat until crisp and golden brown. Stir continuously to prevent the bacon catching and burning. Drain off excess fat and blot the diced bacon on kitchen paper.

Break the eggs into a large bowl and beat to blend

179

well with the cream, a little salt and freshly ground pepper.

Add the bacon to the onion and potato mixture over a fairly gentle heat in the large frying pan, distributing it evenly. When hot, pour the eggs into the pan and tilt it from side to side a few times to let the eggs settle in an even layer. Cook for several minutes, until the omelette has set firmly on the underside. Turn and continue cooking for a few minutes. The outside should be firm, the inside still soft and creamy.

Serve hot or cold, with salad.

———•———

ONION AND BACON PANCAKES

1¼ lb (600 g) onions
4 pancakes (see page 36, half quantity)
1 oz (30 g) butter
4 tbsp sunflower oil
generous pinch caster sugar
8 thick rashers smoked streaky bacon
7 fl oz (200 ml) double cream
salt and pepper

Make the batter for the pancakes ready the day before or at least 6 hours in advance and store, covered, in the refrigerator.

Trim and peel the onions, cut from top to bottom in half and then slice both halves very finely in the same direction. Heat the butter and 3 tbsp of the oil in a wide, non-stick frying pan and cook the onions very gently for 20–25 minutes, stirring frequently. Sprinkle with the sugar halfway through this time, to make them colour well.

Cut the rind off the bacon and cut across the rashers into ¼-in (½-cm) wide strips. Fry over a high heat in a non-stick frying pan, stirring continuously until crisp and golden brown. Drain off all the fat and blot the crisp bacon on kitchen paper before adding it to

the onions. Season sparingly with salt and generously with freshly ground pepper and set aside. Lightly oil a very wide non-stick frying pan (approx. 10 in (25 cm) in diameter) with oil and heat. Cook 4 pancakes.

Heat the onion and bacon mixture over a low heat and correct the seasoning.

Return a pancake to the very wide frying pan, spread a quarter of the onion and bacon mixture over it and heat for about 30 seconds. Fold all four 'sides' of the pancake over the filling to meet in the middle, forming a square envelope. Repeat with the other pancakes and the rest of the filling. Serve on very hot plates.

———•———

PROVENÇAL PANCAKES

2 large onions
2 large yellow peppers
½–1 green chilli pepper (optional)
2 medium-sized ripe tomatoes
1 small clove garlic, peeled and grated
2 tbsp finely chopped parsley
6 fresh basil leaves, torn into small pieces
2 oz (60 g) coarsely grated Gruyère or Emmenthal cheese
scant 2 oz (50 g) coarsely grated Parmesan cheese
7 fl oz (200 ml) pancake batter (see page 36)
5 tbsp sunflower oil
oregano
salt and pepper
Serve with:
Turnip and Rice Purée (see page 193)
Garnish with:
radish flowers (see page 28) or tomato wedges
sprigs of parsley

Make the pancake batter at least 2 hours in advance and chill in the refrigerator (see recipe on page 36). You will need only a quarter of the volume of the original recipe so reduce the ingredients by three-quarters.

Make the filling: peel the onions and slice thinly. Heat 3 tbsp of the oil in a wide, fairly deep frying pan and fry the onions very gently for 10 minutes, stirring frequently. Rinse and dry the peppers, remove their stalk, seeds and white membrane and cut into very thin strips. Add to the onions with the chilli pepper and fry over a moderate heat for 5 minutes, stirring.

Peel the tomatoes and remove their seeds, chop the flesh coarsely and add to the onion and pepper mixture with the garlic. Stir well, cover and simmer gently over a low heat for 15 minutes, stirring now and then. Add a pinch of oregano and season with salt and freshly ground pepper, and cook, stirring continuously, over a moderate heat to allow the liquid to evaporate and the mixture to thicken. Add a little more salt to taste, stir in the parsley and basil and leave to cool to room temperature.

When cold, transfer to a large bowl. Grate both cheeses. Mix with the pepper and onion mixture and stir in the pancake batter. Lightly oil a small non-stick frying pan about 6½ in (16 cm) in diameter and set to heat over a moderate heat for about 2 minutes, when it should be very hot. Ladle a quarter of the pancake batter (about 4 fl oz (120 ml)) into the centre of the frying pan, quickly tipping it from back to front and side to side to distribute the mixture all over the surface of the pan while it is still liquid. Cook for 3–4 minutes, or until set and pale golden brown on the underside, then turn carefully and cook for a further minute.

Keep hot in a warm oven while you cook the other pancakes. Serve without delay, with Turnip and Rice Purée (see page 193). Decorate with radish flowers or tomato wedges and sprigs of parsley.

———— • ————

SPICED VEGETABLES
Makhanwala

3 medium-sized onions, peeled and finely chopped
4 oz (100 g) French beans
4 oz (100 g) peeled, diced potatoes
4 oz (100 g) peeled, diced carrots
4 oz (100 g) cauliflower, cut into small florets
2 oz (50 g) peas
2 oz (50 g) tinned black beans, drained
6 tbsp ghee (clarified butter, see page 36)
4 tbsp coriander leaves, finely chopped
2 tsp mild curry powder
½ tsp ground cumin
1 tsp paprika
9 fl oz (250 ml) single cream
4 tbsp tomato ketchup
salt
For the masala:
4 tbsp grated fresh coconut or creamed coconut
2 tbsp grated fresh ginger
1–2 green chilli peppers, finely chopped
2 large cloves garlic
Garnish with:
2 tbsp chopped coriander leaves
Serve with:
chapatis (see page 81)
rice

Trim, prepare and rinse the vegetables.

Place all the masala ingredients in the food processor and blend well with 3–4 tbsp water.

Heat the ghee in a wide, fairly deep frying pan and fry the onions gently over a low heat for 10 minutes, stirring frequently, until they are soft and pale golden brown. Add the 4 tbsp chopped coriander leaves, the curry powder, cumin and paprika. Cook for a few seconds, stirring, then add the masala mixture, together with 1 tsp salt. Stir well and continue cooking gently for 10 minutes. Add the vegetables and stir so that they are coated and flavoured with the spicy mixture. Add approx. 3½ fl oz (100 ml) water, stir once more, cover and simmer very gently over a low heat for 15–20 minutes. When the vegetables are almost tender but still very crisp, add the cream and ketchup and then continue cooking, uncovered, for about 10 minutes, stirring from time to time, until the vegetables are tender but still firm. Taste, add more salt if needed, then serve piping hot, sprinkled with the 2 tbsp chopped coriander leaves and accompanied by chapatis and plain steamed or boiled rice. Serves 6.

—— • ——

SCRAMBLED EGGS WITH SPRING ONIONS INDIAN STYLE

10 large spring onions
½–1 green chilli pepper
ghee (clarified butter, see page 36)
1½ tsp cumin seeds
2 tbsp concentrated tomato purée
8 eggs
4 fl oz (120 ml) double cream, heated
2 tbsp chopped coriander leaves
salt
Serve with:
chapatis (see page 81), Stuffed Parathas (see page 49) or Spiced Lentils and Spinach (see page 238)

Cut off the root end and remove the outermost layer of the spring onions; slice both the bulb and leaves into thin rings. Remove the stalk and seeds from the chilli pepper and chop finely. Heat 4 tbsp ghee in a wide, non-stick frying pan and gently fry the spring onions and the chopped chilli over a low heat for 12 minutes, stirring frequently.
Toast the cumin seeds for about 30 seconds in a small non-stick frying pan over a fairly high heat,

stirring continuously with a wooden spatula or spoon; when they start to jump about, they are done. Remove from the heat immediately and set aside. Mix the tomato purée with 2 tbsp hot water and stir into the onions over a moderate heat for 1 minute to allow excess moisture to evaporate.
Break the eggs into a large mixing bowl and beat briefly with a pinch of salt. With the heat on very low under the frying pan containing the onions, pour in the beaten eggs; stir until the egg mixture has thickened but is still very creamy (about 1–1½ minutes). Remove from the heat, quickly add the cream (have this ready warmed), the cumin seeds and chopped coriander leaves and stir for about ½–1 minute. Serve at once on very hot plates.

—— • ——

PERSIAN VEGETABLE AND EGG TIMBALE
Kukuye Sabzi

2 large leeks
2 spring onions
1¼ lb (600 g) spinach (net trimmed weight)
1 firm lettuce
2 tbsp finely chopped parsley
28 shelled walnut halves
8 eggs
2 tbsp plain flour
olive or sunflower oil
salt and black peppercorns
For the sauce:
7 fl oz (200 ml) Greek yoghurt
7 fl oz (200 ml) whipping cream
salt
Garnish with:
seeds from ½ ripe pomegranate (optional)
small mint leaves

Remove the stalks from the spinach and reserve them for soups or for braising; you will use only the leaves for this recipe. Wash these very thoroughly, dry in a salad spinner or in clean cloths, chop very finely and place in a large mixing bowl.

Trim, wash and prepare the leek, spring onions and lettuce. Remove the outer, tougher layers or leaves of all these, as only the tender parts are used for this recipe. Cut them all into very thin strips. Mix in the bowl with the spinach and the parsley. Preheat the oven to 325°F (170°C) mark 3. Grease a non-stick ring mould lightly with oil (approx. 10 in (25 cm) outer diameter, 4 in (10 cm) inner diameter).

Chop the walnuts very coarsely. Break the eggs into a bowl and beat them briefly with a little salt and freshly ground black pepper. Sift the flour into the bowl and beat well. Add to the prepared vegetables and mix thoroughly. Carefully ladle into the mould, tap the bottom of the mould on the work surface gently to eliminate any trapped air and cook in the oven for 50 minutes, or until the surface is pale golden brown and the cooked mixture has shrunk away from the sides a little. Test by inserting a thin skewer deep into the mixture: it should come out clean and dry. Take out of the oven and leave to stand for 15 minutes before unmoulding on to a serving plate.

While the mould is cooking, make the sauce. Beat the cream stiffly, fold in the yoghurt and add a pinch of salt. Pour some of the sauce into a small bowl and place in the central well of the mould; alternatively spoon in enough sauce to fill this central cavity. The remaining sauce can be handed round separately in another small bowl. Arrange the pomegranate seeds and the mint leaves decoratively on top of the creamy sauce, both in the mould and in the extra bowl. Serve this dish warm or at room temperature. Serves 6.

— • —

LEEK AND CHEESE PIE

1¾ lb (800 g) leeks, (net cleaned, trimmed weight)
approx. 14 fl oz (400 ml) milk
1 tbsp cornflour or potato flour
3½ fl oz (100 ml) double cream
pinch grated nutmeg
4½ oz (130 g) grated hard cheese
2 oz (50 g) butter, softened at room temperature
salt and pepper

Bring plenty of salted water to the boil in a very large saucepan. Trim and wash the leeks, removing the tough, outer layer and the ends of the green leaves. Keep them whole but cut down into quarters from the top, green end to within about 3 in (7 cm) of the root end to enable you to wash them thoroughly. Boil for 20 minutes and drain. Preheat the oven to 400°F (200°C) mark 6.

Bring the milk slowly to the boil in a small saucepan. Mix the cornflour or potato flour with 2 tbsp cold water and stir into the hot milk off the heat. Return to the heat and continue stirring for about ½–1 minute as the milk thickens. Remove from the heat and beat in the cream a little at a time using a balloon whisk. (This is not a thick sauce.) Season with salt, pepper and nutmeg.

Use a little of the butter to lightly grease a shallow ovenproof dish large enough to accommodate all the leeks. Slice the leeks across, cutting them in half. Arrange in a single layer in the greased dish alternating a green half with a white half. Cover with the white sauce. Sprinkle with the cheese and dot with flakes of the remaining butter; bake in the oven for 25 minutes and finish by placing under a very hot grill for 5 minutes to brown. Serve at once.

— • —

STEAMED VEGETABLES WITH CURRY SAUCE

4 leeks
2 waxy potatoes
2 courgettes

183

2 small turnips
2 large bulbs fennel
salt
For the curry sauce *(makes 18 fl oz (500 ml)):*
9 oz (250 g) creamed coconut
1 large onion
½ green chilli pepper
3 tbsp ghee (clarified butter, see page 36)
3½ fl oz (100 ml) chicken stock (see page 37)
1 tbsp plain flour
1 piece fresh ginger, peeled and grated
3 tbsp mild curry powder
2 tbsp cider vinegar or wine vinegar
approx. 1 tsp fresh lime juice
2 tbsp chopped coriander leaves
salt and pepper
Garnish with:
4–6 sprigs coriander

Make the coconut milk for the curry sauce: break up the block of creamed coconut into fairly small pieces and place in a large bowl; add 14 fl oz (400 ml) boiling water and leave to stand for 30 minutes. Meanwhile, prepare all the vegetables, cut into fairly small pieces and arrange in the steamer. Strain through a clean linen napkin or two layers of muslin cloth, twisting the ends of the cloth round tightly to squeeze out all the moisture. Discard the solid material left inside the cloth and heat the coconut milk.

Chop the onion finely; do likewise with the chilli pepper, discarding the seeds. Cook the onion in the ghee over a low heat in a wide, heavy-bottomed saucepan for 10 minutes while stirring with a wooden spoon. Add the chilli pepper and the ginger and continue frying gently for a further 5 minutes. Stir in the flour and cook for a few seconds; mix the curry powder with the vinegar and stir into the onion mixture. Start steaming the vegetables at this point; they will take approx. 12–15 minutes. Remove the curry mixture from the heat and pour in the scalding

hot coconut milk all at once while beating continuously with a balloon whisk to prevent lumps forming. Return to a low heat, beat in the stock and keep beating as the mixture simmers until it acquires a velvety consistency.

Use a hand-held electric beater to beat the hot sauce until it is smooth. Season with salt and pepper and add a little lime juice or lemon juice to taste. Just before serving add the chopped coriander.

Transfer the steamed vegetables to heated plates, cover with the curry sauce and garnish each serving with a sprig of coriander.

———— • ————

RUSSIAN CARROT MOULD

1¼ lb (600 g) young tender carrots
6 eggs
2 tbsp chopped fresh dill
3 oz (80 g) butter, 2½ oz (70 g) of which melted
2½ oz (70 g) fine fresh breadcrumbs
2 tbsp fine dry breadcrumbs
nutmeg
salt and pepper

Preheat the oven to 350°F (180°C) mark 4. Use ½ oz (10 g) of the butter to grease a 2-pint (1.2-litre) soufflé dish; sprinkle all over the inside with fine dry breadcrumbs. Trim and peel the carrots. (The fresher the carrots, the more flavour they will have.) Cut into small pieces and boil in a large pan of salted water for about 30 minutes or until very tender. Drain well and purée in a food mill with a fine strainer fitted. Transfer to a large bowl, stir in the 6 egg yolks, the dill, salt, freshly ground white pepper to taste and a pinch of nutmeg. Stir, then add the melted butter and the breadcrumbs and mix well. Beat the egg whites with a pinch of salt until stiff but not at all dry or grainy and fold into the carrot mixture. Tip the mixture into the prepared soufflé dish, smoothing the top level with a palette knife. Place in a roasting tin, pour sufficient boiling water into the tin to come about halfway to two-thirds of the way up the sides and cook for 40 minutes.

Leave to stand and set for a few minutes before turning out the carrot mould on to a hot serving dish.

———•———

SCORZONERA GREEK STYLE

2 lb (900 g) scorzonera (black salsify)

approx. 14 fl oz (400 ml) chicken stock (see page 37)

juice of 2 lemons

1 medium-sized onion

2 tbsp chopped parsley

4 tbsp extra-virgin olive oil

1½ tbsp plain flour

3 egg yolks

salt and pepper

Set the stock to heat to boiling. Stir the juice of 1 lemon into a large bowl of cold water. Trim, wash and peel the scorzonera one root at a time. As each root is peeled, cut it lengthwise into thick slices; cut each of these into 1½-in (4-cm) lengths and remove any of the hard, woody central section from each piece with a sharp knife. Drop these into the bowl of acidulated water to prevent them discolouring.

Chop the onion finely. Heat the oil in a large, heavy-bottomed saucepan and fry the onion gently with the chopped parsley for a few minutes, until tender. Drain the scorzonera well, add to the pan and sauté for 5 minutes. Add a pinch of salt, sprinkle with the flour and continue stirring for 1 minute to cook the flour and distribute it evenly. Add approx. 16 fl oz (450 ml) of the boiling hot stock and stir well. Reduce the heat to low, cover and simmer for 10 minutes, adding a little more hot stock when needed.

When the scorzonera is tender, turn down the heat as low as possible and make a thickening liaison by beating the egg yolks with the juice of the second lemon and 2 tbsp of the hot stock in a small bowl. Sprinkle over the scorzonera and continue cooking, stirring continuously, over a very low heat for 1–2 minutes. The egg yolk will thicken the sauce a little and make it slightly creamy; as soon as it does, remove from the heat. This is not meant to be a thick sauce. Continue stirring for half a minute off the heat, correct the seasoning and serve without delay.

———•———

SALSIFY POLISH STYLE

1¾ lb (800 g) salsify (also known as oyster plant)

vegetable stock (see page 38)

juice of 1 lemon

4 hard-boiled eggs

3 tbsp finely chopped parsley

10 tbsp ghee (clarified butter, see page 37)

3 tbsp fine dry white breadcrumbs

salt and pepper

Make the vegetable stock and set it to heat. Have a large bowl of cold water acidulated with the lemon juice standing ready and drop each piece of salsify into it as soon as it is peeled to prevent discoloration. Prepare the vegetable in the same way as described for scorzonera in the preceding recipe. Drain and boil in the stock for about 15 minutes, or until tender. Drain and cut across the pieces of salsify into slices.

Transfer to a very hot serving dish and sprinkle with the finely chopped eggs and parsley. Keep hot while you heat the ghee until it is very hot and starting to deepen in colour; add the breadcrumbs, a pinch of salt and a little freshly ground pepper. Stir well, then sprinkle all over the salsify and serve at once.

NEAPOLITAN POTATO CAKE
Gattò di Patate

3¼–3½ lb (1½ kg) floury potatoes

9–12 fl oz (250–300 ml) milk

3 oz (80 g) butter

2 eggs

3 oz (80 g) grated Parmesan cheese

1 lemon

5½ oz (160 g) smoked cheese, diced

4 oz (120 g) Mortadella sausage or ham, sliced

15 basil leaves, torn into small pieces

1 large (or 2 standard size) mozzarella cheeses, sliced

2 oz (60 g) thinly sliced (Neapolitan type) peppery sausage

fine dry breadcrumbs

salt and pepper

Wash the potatoes and steam until tender. Spear them with a carving fork and peel while still boiling hot. Push them through the potato ricer while still hot; repeat the ricing operation, and then finally push them through a fine sieve into a large bowl.

Preheat the oven to 350°F (180°C) mark 4. Heat the milk together with 2 oz (60 g) of the butter and a little salt and freshly ground pepper until scalding. Stir most of the milk into the sieved potato; work in the eggs, Parmesan cheese and the finely grated rind of the lemon. The mixture should be fairly soft; add the rest of the milk if necessary. Stir this mixture thoroughly for 2 minutes, then add the smoked cheese, the chopped Mortadella or ham and the basil. Grease a cake tin about 9½ in (24 cm) in diameter and at least 2 in (5 cm) deep. Place a quarter of the mixture in the tin and press out flat into an even layer with the palm of your hand. Spread out 6 slices of mozzarella on the surface and cover with another quarter of the mixture. Repeat these last two layers. Spread the sausage slices on top. Sprinkle breadcrumbs over the surface and dot small flakes of the remaining butter on top. Bake for about 30 minutes, browning under a very hot grill for a final 5 minutes to form a crisp topping. Serves 8–10.

———— • ————

PAN-FRIED POTATOES, PEPPERS AND BEEF

2 large waxy potatoes

2 large yellow peppers

4 cloves garlic, peeled

1 red chilli pepper

generous pinch oregano

2 tbsp finely chopped parsley

14 oz (400 g) thickly sliced beef fillet

4 tbsp extra-virgin olive oil

2 tbsp light olive oil

salt and pepper

Peel the potatoes and cut into small cubes. Blanch in a large pan of boiling salted water for 5 minutes, drain and leave to cool. Cut the peppers in half, remove the stalks, seeds and white membrane and slice into wide strips. Cut these into approx. 1½-in (4-cm) lengths.

Heat the extra-virgin olive oil in a large frying pan or wok and sauté the garlic cloves, crushed with the flat of a knife but still left whole, together with the peppers. Add the finely chopped chilli pepper, the oregano and half the parsley. Cook over a fairly high heat for 2 minutes. Sprinkle with a pinch of salt, sauté for another minute and add the potatoes. Stir, cover the pan and cook over a low heat for 10 minutes, stirring every few minutes. Cut the beef fillet into small cubes and season these with salt and freshly ground black pepper. Heat the light olive oil over a high heat in a large, non-stick frying pan until very hot; add the beef and stir-fry for 1 minute. Add to the potatoes, stir for a few seconds and then take off the heat. Sprinkle with the remaining parsley and serve at once on heated individual plates.

VEGETABLE AND CHEESE BAKE

2½ lb (1.1 kg) waxy potatoes

2 large onions

1¼ lb (600 g) tinned tomatoes, drained

3 oz (80 g) grated hard cheese

extra-virgin olive oil

generous pinch oregano

salt and pepper

Serve with:

nut roasts

grilled meat, fish or poultry

Preheat the oven to 325°F (170°C) mark 3. Peel the potatoes and cut into slices about ⅜ in (1 cm) thick; as you cut the slices, put them in a large bowl of cold water to prevent them turning brown. Peel the onions and slice very thinly.

Drain the tomatoes, reserving the juice. Cut the tomatoes open and spoon all their seeds into the sieve; drain off any more juice they release and keep it. Discard the seeds. Chop the flesh coarsely.

Grease a wide, shallow ovenproof dish with oil. Drain the potatoes and transfer to the oven dish together with the onions and tomatoes. Sprinkle with the reserved tomato juice. Season with salt, freshly ground pepper and sprinkle with the oregano and the grated cheese. Finally, moisten with about 9 fl oz (250 ml) boiling water and sprinkle with 2 tbsp olive oil. Bake for 1¼–1½ hours or until the potatoes are very tender and the top is crisp and golden. Allow to stand for 10 minutes before serving.

———•———

POTATO AND ARTICHOKE PIE

4 large waxy potatoes

5 artichoke hearts (see method)

1 lemon

4 large cloves garlic

2 tbsp finely chopped parsley

extra-virgin olive oil

salt and pepper

Serve with:

sausages

braised onions

Preheat the oven to 325°F (170°C) mark 3. Peel the potatoes, cut into fairly thin slices, dropping them into a large bowl of cold water as you do so.

If you use very tender, young artichokes prepare as described on page 18, cut each artichoke heart lengthwise in quarters, cut each quarter lengthwise in half. Drop into a large bowl of cold water acidulated with the lemon juice as you prepare them to prevent discoloration. If using larger, less tender varieties of artichoke, use 10, strip off all the leaves and remove the choke from the bottom.

Drain the vegetables and spread out evenly in a large, shallow ovenproof dish. Season with salt and pepper; sprinkle with 4 tbsp olive oil, the crushed garlic and 9 fl oz (250 ml) boiling water. Cook for 1 hour or until tender. Alternatively, sauté the vegetables in the oil with the garlic in a large frying pan for about 8 minutes, then add the other ingredients, including the water, cover, reduce the heat and simmer for about 20 minutes, adding a little more hot water as required.

———•———

POTATO SOUFFLÉ WITH ARTICHOKE SAUCE

1¾ lb (800 g) floury potatoes

3 oz (80 g) butter

¼ lb (120 g) finely grated Gruyère or Emmenthal cheese

187

4 eggs
small pinch cream of tartar or 2–3 drops lemon juice
nutmeg
salt and pepper
Serve with:
artichoke sauce (see page 77)

1 2-in (5-cm) piece horseradish root (see method)
1 green apple
3½ fl oz (100 ml) low-fat plain yoghurt
3½ fl oz (100 ml) double cream
1 tbsp fresh lime or lemon juice
1 tsp cane sugar
cider vinegar
nutmeg
salt and pepper

Grease a 2-pint (1.2-litre) or slightly larger soufflé dish with ½ oz (10 g) of the butter. Preheat the oven to 350°F (180°C) mark 4. Scrub the potatoes under running cold water and steam until tender. Spear the potatoes with a carving fork and peel while still boiling hot; put through a potato ricer twice, or push through a fine sieve. Transfer the mashed potatoes to a large mixing bowl. Melt the remaining butter, and add it to the mashed potatoes, together with the cheese and the egg yolks, adding a pinch each of salt, pepper and grated nutmeg.

Beat the egg whites with a pinch of salt, a small pinch of cream of tartar (or a few drops lemon juice) until they are stiff; fold in the potato and cheese mixture. Pour into the prepared soufflé dish and bake for 35–40 minutes, when the soufflé should have risen well in the dish and turned golden brown on top. Serve at once with the artichoke sauce: make this while the soufflé is cooking.

—— • ——

SWEET POTATO PANCAKES WITH HORSERADISH AND APPLE SAUCE

1¼ lb (600 g) sweet potatoes
2 oz (50 g) plain flour
1 small onion, finely chopped (optional)
2 eggs
1 tbsp ghee (clarified butter, see page 36)
nutmeg
salt and pepper
For the horseradish and apple sauce:

Heat plenty of salted water in a large saucepan. Peel the potatoes, grate them coarsely with a grater or in the food processor and blanch them in the boiling salted water for 5 minutes. Drain, transfer to a clean cloth or linen napkin, gather up the edges and twist them round tightly to force out all excess moisture. Mix the potato in a large bowl with the flour, onion and the lightly beaten eggs, seasoned with salt, pepper and a pinch of nutmeg. Cover the bowl and chill this batter in the refrigerator.

Make the horseradish and apple sauce: peel the horseradish (wear sunglasses as it will make your eyes water and sting; alternatively, use vacuum packs or jars of ready grated moist horseradish which is not so strong). Peel the apple, cut into pieces and process in a food processor with the horseradish until smooth and creamy. Transfer to a bowl, stir in the lime or lemon juice, the sugar, ½ tsp vinegar and a pinch each of salt and freshly ground pepper. Mix in the yoghurt and cream and add more salt to taste.

Make the pancakes: heat 1 tbsp clarified butter in a small non-stick frying pan (approx. 6 in (15 cm) in diameter) and when very hot, add a quarter of the batter, pressing the mixture level with a non-stick spatula before it sets. Fry over a moderate heat for 3 minutes, then turn the pancake and cook for another 2 minutes, when it should be browned. Drain and keep hot.

These pancakes look more like large fritters or thin potato cakes. Serve as soon as the last one is cooked, handing round the horseradish and apple sauce separately.

GREEK POTATO FRICADELLES
Patatokeftédes

2¼ lb (1 kg) floury potatoes

3 eggs

1 small onion, grated

3½ oz (100 g) grated hard cheese

3 tbsp finely chopped parsley

light olive oil

fine dry breadcrumbs

nutmeg

salt and pepper

Garnish with:

parsley

lemon wedges

Serve with:

beetroot salad

Scrub the potatoes clean under running cold water and steam until very tender; spear with a carving fork and peel while still boiling hot. Put them through a potato ricer twice (or push through a fine sieve) and place in a large mixing bowl. Work in 1 whole egg and 1 yolk (reserve the white from this second egg). Mix in the onion, cheese, parsley and a generous pinch each of salt, pepper and grated nutmeg.

Break off pieces of the dough mixture and shape between your palms to the size of a small egg; press gently to flatten slightly. Lightly beat 2 egg whites (you will have 1 yolk left over, which you can reserve for use in another recipe); spread out about 4 oz (120 g) breadcrumbs in a large plate. Dip the fricadelles in the egg whites to moisten all over and then coat thoroughly with the breadcrumbs. Arrange on a plate or chopping board, cover with cling film and chill in the refrigerator for at least 1 hour.

When it is time to cook the fricadelles, heat plenty of oil in a deep-fryer to 350°F (180°C) mark 4 and fry them 2 or 3 at a time until crisp and golden brown.

Place in a single layer on kitchen paper, uncovered, to keep hot in the oven while you finish frying the rest. Serve without delay on a heated oval platter, garnished with sprigs of parsley and lemon wedges, accompanied by a beetroot salad dressed with oil, wine vinegar, salt, freshly ground pepper and sprinkled with finely chopped parsley.

———•———

VEGETABLES WITH POTATO STUFFING

1 lb (500 g) floury potatoes

12 large, freshly picked pumpkin or courgette flowers (see method)

1 very large onion

4 young courgettes

2 eggs

3 oz (80 g) freshly grated hard cheese

1 tsp fresh marjoram leaves

1 clove garlic, peeled

fine dry breadcrumbs

thyme

4 tbsp olive oil

butter

salt and black pepper

Scrub the potatoes under running cold water and steam until tender; prepare the pumpkin flowers carefully by trimming the stalk to within ⅜ in (1 cm) of the flower, remove the pistils delicately and rinse briefly in cold water; spread out to dry on a clean cloth. Peel the onion, trim off the ends and make deep cross cuts in both top and bottom. Trim the ends off the courgettes and rinse. Blanch the onion and courgettes in a large pan of boiling salted water for 10 minutes (take the courgettes out after 5–7 minutes if they are very thin); drain and leave to cool.

189

Preheat the oven to 350°F (180°C) mark 4. When the potatoes are very tender, spear them with a carving fork and peel while still very hot; push through a potato ricer or sieve into a large mixing bowl. Stir in scant 1 tbsp butter, 2 tbsp oil, 2 eggs and two-thirds of the grated cheese. Mix the remaining grated cheese with 3 tbsp dry breadcrumbs and set aside. Chop the marjoram leaves finely with the peeled garlic and stir into the potato mixture together with a generous pinch of thyme, a pinch of salt and plenty of freshly ground black pepper. Mix very thoroughly, adding a little more salt if wished.

Lightly oil a large baking sheet. Cut the courgettes lengthwise in half, scoop out some of the flesh, leaving a layer of flesh in place next to the skin. Slice the onion lengthwise in half and separate the layers to form shallow saucers. Fill the pumpkin or courgette flowers carefully with 2 or 3 tsp of the mixture; do not overfill. Pinch the ends of the flowers gently to seal. Sprinkle a little salt over the partially hollowed courgettes and the onion pieces and fill with stuffing mixture; again, do not overfill. Transfer to the baking sheet, sprinkle each stuffed item with a little olive oil and with the cheese and breadcrumb mixture. Bake for 15–20 minutes, placing the baking sheet under a very hot grill for the last 2–3 minutes to brown the topping. Serve at once, decorating the larger vegetables with a small sprig of fresh basil.

This mixture makes a good stuffing for most vegetables.

STUFFED VEGETABLE SELECTION

3 large waxy potatoes
4 courgettes
2 long aubergines
1 large onion
1 large red pepper
5 oz (150 g) grated Gruyère or Emmenthal cheese
5 oz (150 g) lean cooked ham, diced
1 whole egg
2 tbsp béchamel sauce (see page 36)
2 large ripe tomatoes
approx. 4 oz (120 g) fine dry breadcrumbs
olive oil or sunflower oil
2 oz (60 g) butter
salt and pepper

Start the preparation several hours in advance or the day before you plan to serve this dish by blanching the vegetables and preparing the stuffing. Bring a very large saucepan of salted water to the boil. Trim, wash and slice the courgettes and aubergines lengthwise in half. Peel the onion and cut into quarters. Add all these to the boiling water; after 5 minutes add the rinsed, whole pepper and continue blanching for a further 5 minutes. Drain.

As soon as they are cool enough to handle, scoop out most of the flesh of the courgettes and aubergines, leaving only a thin layer next to the skin, and spoon it into a large mixing bowl. Spread out the hollowed courgettes and aubergines upside down on a slightly tilted chopping board to finish draining.

Process the grated cheese, the ham and the vegetable pulp in the food processor until smooth and evenly blended; transfer this back to the large mixing bowl and stir in the egg and the béchamel sauce. Season with salt and pepper. Chill in the refrigerator for at least 2 hours.

Preheat the oven to 325°F (170°C) mark 3. Grease 1 or 2 fairly shallow baking trays with oil. Peel the potatoes and slice very thinly. Spread them out in a single layer on the baking trays, brush lightly with a little more oil and season with a little salt and freshly ground pepper. Fill the courgettes and the aubergines with some of the chilled stuffing mixture. Cut the pepper lengthwise into quarters, remove the stalk, seeds and white membrane and place some stuffing mixture in each concave quarter, smoothing it neatly.

Separate the layers of onion if they have not already come apart when blanching; fill each concave slice with some stuffing mixture. Cut the tomatoes horizontally in half, scoop out the seeds and some of the flesh and stuff them with the remaining mixture. Place all the vegetables, filling uppermost, on top of the potato slices, arranging them so that the colours alternate attractively, sprinkle with breadcrumbs, place a sliver of butter on top of each and bake for about 50 minutes until golden brown and crisp.

———— • ————

Boiled or Steamed Vegetables with Guasacaca Sauce

3 medium-sized waxy potatoes
4 leeks
4 small tender turnips
4 carrots
1¾ pints (1 litre) chicken stock (see page 37)
For the guasacaca sauce:
2 ripe avocado pears
1 ripe tomato
1 finely chopped hard-boiled egg
1 tbsp chopped coriander leaves
½ tbsp finely chopped parsley
½ green chilli pepper, seeds removed
1 tbsp red wine vinegar
½ tsp pili-pili hot relish (see page 113) or pinch cayenne pepper
3 tbsp olive oil
salt and pepper

Heat the stock. Trim, wash and dry all the vegetables, peeling where necessary; cut into fairly large pieces and boil gently in the stock for 20–30 minutes or until they are tender. Prepare the sauce. Peel the avocadoes, take out the stone, and mash the flesh to a purée.

Peel the tomato (blanch first if necessary), remove all the seeds and any tough parts; dice the flesh and stir into the avocado together with the chopped egg, coriander, parsley and the finely chopped green chilli pepper. Mix in the olive oil, vinegar, pili-pili relish or cayenne pepper, and salt and pepper to taste. Serve the drained vegetables hot, handing round the sauce separately.

Avocado pear flesh discolours within a fairly short time, so do not prepare more than 30 minutes before serving.

———— • ————

Braised Onions and Rice
Soubise

2¼–2½ lb (1 kg) strong onions
4 oz (120 g) risotto rice, parboiled
3 oz (90 g) butter
3 fl oz (80 ml) double or single cream
1 oz (30 g) grated Gruyère or Emmenthal cheese
1½ tbsp finely chopped parsley
salt and pepper
Serve with:
peppered beef fillet (steak au poivre)
braised globe artichokes
Chicory with Lemon Flemish Style (see page 86)

Preheat the oven to 300°F (150°C) mark 2. Bring a large saucepan of salted water to the boil, add the rice and boil briskly for 4 minutes, then drain. Peel and thinly slice the onions.

Melt 2 oz (60 g) of the butter in a large, fireproof casserole dish; add the onions and stir well to coat with the butter. Add the rice, ½ tsp salt, a little freshly ground white pepper and stir well. Cover tightly and cook in the oven for 1 hour, stirring now and then. The onions and rice should be tender and very pale golden brown when done. Add a little more salt and

191

pepper if necessary, followed by the cream, cheese and the parsley. This is a delicious accompaniment for peppered beef fillet (steak *au poivre*); braised artichokes or braised chicory Flemish style would complete a main course.

———•———

SWEET–SOUR BABY ONIONS

14 oz (400 g) baby onions, peeled

2 small pieces of orange rind

14 fl oz (400 ml) red wine vinegar

8 tbsp cane sugar (white, demerara or soft light brown)

salt

Place half the onions in a single layer in a wide, fairly shallow flameproof casserole dish, half in another, similar casserole dish. Sprinkle both with a little salt and add 5 fl oz (150 ml) water, 7 fl oz (200 ml) vinegar and a piece of orange rind to both batches. Heat to boiling point, cover and cook over a low heat for 10 minutes.
Sprinkle half the sugar over each batch of onions, stir, replace the lids and simmer very gently for 10 minutes more or until the liquid has almost completely evaporated and the surface of the onions is glazed and glossy.

———•———

TURKISH RICE WITH LEEKS
Zeytinyagli Pirasa

1 lb (900 g) leeks

1 medium-sized onion, finely chopped

4 oz (130 g) long-grain rice, parboiled

3 tbsp olive oil

1 tbsp plain flour

½ tsp caster sugar

salt

Garnish with:

1 lemon

Prepare the leeks; trim off the roots and remove the tough, outer layers. Cut off most of the green part, leaving about 2 in (5 cm) in place. Wash very thoroughly. Cut lengthwise in quarters, then slice each length into 1–1¼-in (3-cm) sections. Cook the chopped onion in the oil over a low heat in a very large frying pan or wide saucepan for about 10 minutes, stirring frequently until it is tender and wilted but not at all browned. Stir in the flour, ½–1 tsp salt and ½ tsp caster sugar and continue stirring for 2 minutes; 1–1½ tbsp tomato purée can be added at this point for extra flavour if wished. Gradually add 14 fl oz (400 ml) hot water, stirring continuously with a wooden spoon or beating with a balloon whisk. Increase the heat to moderately high and keep stirring or beating as the mixture comes to the boil; add the leeks. Mix thoroughly, turn down the heat to very low, cover tightly and simmer gently for 10 minutes. When this time is up, sprinkle in the rice, stir well, cover tightly again and cook for a further 15 minutes, until the rice is tender. Garnish with lemon wedges.

———•———

CARROTS WITH CREAM AND HERBS

1¾ lb (800 g) carrots

2 finely chopped shallots

2 oz (50 g) butter

7 fl oz (200 ml) crème fraîche or single cream

2 pinches caster sugar

generous pinch oregano

2 tbsp chopped fresh basil

salt and pepper

Serve with:

Spinach Soufflé (see page 78)

poached salmon

Wash, trim and peel the carrots. Cut them into large matchstick strips. Heat the butter gently in a very wide frying pan and cook the shallots over a very low heat, stirring frequently, for 10 minutes, until wilted and tender. Add the carrots, sprinkle with a little salt, freshly ground pepper and 2 pinches of caster sugar. Stir over a gentle heat for 1 minute, then cover and sweat slowly for about 30 minutes or until the carrots are tender but still crisp, stirring often and adding a very little hot water when necessary.

Add a generous pinch of oregano, stir over a moderately high heat for a few seconds, then add the cream. Reduce the heat and stir for 2–3 minutes. Remove from the heat and add a little more salt if wished. Serve in a heated serving dish with a sprinkling of chopped fresh basil leaves. Serve at once, before the basil wilts and while the carrots are piping hot.

———•———

DEEP-FRIED SCORZONERA

1¾ lb (800 g) scorzonera

1 egg yolk

juice of 1 lemon

4½ oz (125 g) plain flour, sifted

sunflower oil

salt

Make the batter: beat the egg yolk in a large bowl with ½ tsp salt and 7 fl oz (200 ml) iced water. Sift in

the flour and stir just enough to get rid of any lumps. Cover and chill in the refrigerator.

Bring a large saucepan of salted water to the boil; add half the lemon juice. While it is heating, peel the scorzonera and cut into pieces approx. 1½ in (4 cm) long. As each piece is prepared, drop it into a large bowl of cold water acidulated with the remaining lemon juice to prevent discoloration. When the water boils, drain the scorzonera pieces and boil for 18 minutes or until tender but still firm. Drain well and leave to cool a little. Cut each piece in half lengthwise and remove the tough, woody central section.

Heat plenty of oil in a deep-fryer to 350°F (180°C) and deep-fry a few pieces of the scorzonera at a time, until they are crisp and golden; make sure the temperature of the oil remains constant. Remove each batch from the oil with a slotted spoon and finish draining on kitchen paper; keep hot while you finish the deep-frying. Serve immediately.

———•———

TURNIP AND RICE PURÉE

3 large turnips

8 oz (230 g) long-grain rice

18 fl oz (500 ml) milk

2 large cloves garlic

½ tsp fresh thyme leaves or pinch dried thyme

5 tbsp single cream

2 tbsp finely chopped parsley

1 oz (30 g) butter

salt and white pepper

Serve with:

Mushrooms Bordeaux Style (see page 264)

Peas with Bacon and Basil (see page 234)

Peel and coarsely chop the turnips. Bring the milk slowly to the boil in a large, heavy-bottomed, fireproof casserole dish, then sprinkle in the rice, add the butter, the peeled, whole garlic cloves, the thyme and ½ tsp of salt. Mix well. Simmer for 10 minutes, stirring every few minutes. Add the turnips and a little more milk if necessary to cover them. Cover, reduce the heat as low as possible and simmer for 15 minutes, stirring at intervals. When the turnips are very tender and nearly all the milk has been absorbed, remove from the heat and beat with the hand-held electric beater until the mixture forms a smooth purée (or put through the vegetable mill with a fine-gauge disc). These first stages can be completed several hours in advance.

When it is time to finish cooking the purée, reheat it over a low heat without a lid on, stirring now and then until it has thickened further. Turn off the heat, stir in the cream, a little freshly ground white pepper and the parsley. Add a little more salt if needed and serve.

---

1 tbsp oil in a wide, non-stick frying pan and as soon as the butter has stopped foaming, sauté half the root vegetables over a fairly high heat for 3–4 minutes until pale golden brown all over, stirring with a wooden spatula.

Use a slotted spoon to transfer this first batch to a bowl. Add a little more butter and oil to the frying pan, and when very hot, sauté the second batch in the same way. Add the first batch to the pan with the clove of garlic (crushed but still in one piece) and a little salt and freshly ground pepper. Reduce the heat to very low, cover and sweat gently for 8–10 minutes, or until tender but still with a little bite left; stir every 2–3 minutes.

Remove from the heat, add a little more salt and pepper if needed; sprinkle with the parsley and serve at once.

BABY TURNIPS SAUTÉED WITH GARLIC AND PARSLEY

2¾ lb (1.2 kg) very small baby turnips
juice of 1 lemon
butter
sunflower oil
1 large clove garlic
3 tbsp finely chopped parsley
salt and pepper
Serve with:
Peas with Bacon and Basil (see page 234)
Mushroom and Herb Omelette (see page 261)

Trim off the ends of the baby turnips and peel with the potato peeler, dropping them immediately into a large bowl of cold water to which the lemon juice has been added, to prevent discoloration. Drain and dry, cut each root from top to bottom into quarters and then into eighths. Heat 1 oz (30 g) butter and

RADISHES WITH CREAM, CORIANDER AND POMEGRANATE SAUCE

3 small bunches small radishes
2 large shallots, finely chopped
1 oz (30 g) butter
3 tbsp raspberry vinegar
10 oz (300 ml) crème fraîche or double cream
scant 2 tbsp chopped fresh coriander leaves
seeds from ⅛ of a pomegranate
salt and pepper
Serve with:
Rösti Potatoes (see page 195), veal dishes or vegetable omelettes

Trim and wash the radishes; cut them into round slices. Heat the butter in a wide, fairly shallow flameproof casserole dish and fry the shallots very gently for about 10 minutes until wilted and tender but not at all browned. Add the radishes, turn up the heat a little and fry for 1 minute, stirring with a wooden spatula. Season lightly with salt and pepper. Sprinkle with the vinegar and continue cooking until the liquid has completely evaporated.

Pour the cream all over the radishes, stir, simmer gently for 2–3 minutes, stirring now and then. Take off the heat, add a little more salt and pepper if necessary and sprinkle with the coriander and pomegranate seeds. Serve straight from the dish.

———•———

BEETROOT, TOMATO AND CUCUMBER SALAD

1 large cooked beetroot

2 firm medium-sized tomatoes

1 large or 2 small cucumbers

4 tbsp grated fresh coconut flesh

2 tbsp chopped fresh coriander leaves

½ green chilli, finely chopped

4 tbsp chopped unsalted peanuts

18 fl oz (500 ml) chilled natural yoghurt

1 tsp caster sugar

1 tbsp sunflower oil

1 tbsp cumin seeds

salt

Serve with:

chapatis (see page 81)

Peel and dice the beetroot and the cucumber. Peel the tomatoes, remove the seeds and dice. Place all the prepared vegetables in a large salad bowl, sprinkle with a pinch of salt and with the sugar and gently stir in the coconut, coriander, chilli pepper, peanuts and yoghurt. Add a little more salt if wished. Heat the oil in a small, non-stick frying pan and when very hot, add the cumin seeds. Fry for 2–3 seconds; as soon as they start to jump about in the pan, take off the heat and add to the salad. Stir carefully once more and serve with chapatis as a delicious, refreshing accompaniment to almost any Indian main course.

———•———

RÖSTI POTATOES

6 large waxy potatoes

1 oz (30 g) butter

salt

Serve with:

fried, poached or baked eggs with crispy bacon or cheese omelettes

Peel the potatoes and chop coarsely in the food processor (or grate using the large-gauge side of a grater). Place in a large mixing bowl and season with a little salt and pepper.

Place a quarter of the butter in two wide, non-stick frying pans; heat and when it starts to sizzle and the foam has disappeared, spread out half the potato in one and half in the other, pressing down well with a non-stick spatula to make the potato cakes firm and evenly spread out. Fry over a moderately high heat for 4 minutes or until the edges have browned well and turned crisp; use plates to help you turn the potato cakes: slide each one in turn on to a plate, place another plate on top, turn upside down; you will then be able to slide the potato cake back into its frying pan. While you have the potato cakes on the plates, divide the remaining butter in half and add to each frying pan; when very hot, slide in the upturned potato cakes. Cook for another 4 minutes, serving as soon as the cakes are evenly browned and crisp. To serve 4, cut each flat potato cake neatly in half. Fried eggs and crispy bacon go well with these potato cakes, as do cheese omelettes.

———•———

Rösti Potatoes with Shallots

6 large waxy potatoes, coarsely grated

8 tbsp finely chopped shallot or mild onion

1 oz (30 g) butter

salt and pepper

The method is the same as for the preceding Rösti recipe on page 195 but, before adding the grated potato to the butter in the frying pans, gently fry half the shallots in each frying pan over a low heat until soft but not at all browned. Press an even layer of the potato on top of the shallot and then proceed as directed in the previous recipe.

·•·

Chicory with Raspberry Vinegar and Cream

2 shallots, finely chopped

4 heads red Treviso chicory (see method) or common chicory

1 oz (30 g) butter

2 tbsp raspberry vinegar

7 fl oz (200 ml) crème fraîche or single cream

salt and pepper

Treviso red chicory is fairly similar in shape to the more common greenish white chicory but it is longer and has a deep reddish pink tinge to the leaf tips and sides. It is still not common in this country. Do not substitute the darker, curlier round red radicchio as this is very bitter when cooked; use ordinary chicory instead. Cut off the solid base of each, wash the leaves, dry in the centrifugal salad spinner and cut into thin strips. Sweat the shallots in the butter in a large, heavy-bottomed stainless steel or enamelled saucepan. Add the prepared vege-

table, turn up the heat slightly and fry gently, stirring with a wooden spoon, for 2 minutes.

Sprinkle with the raspberry vinegar and keep stirring over the heat until it has completely evaporated. Season lightly with salt and pepper, add the cream and mix well. Cover and cook gently over a low heat for 4 minutes. Taste, adjust the seasoning and serve.

·•·

Macedoine of Vegetables with Thyme

5 large waxy potatoes

7 large carrots

7 large courgettes

1 large shallot or mild red onion, finely chopped

2 tbsp clarified butter (see page 36)

1 sprig thyme

salt and pepper

Peel the potatoes and place in a large saucepan with enough cold water to cover; set them aside while you trim, peel, rinse and dry the carrots and courgettes. Heat a large pan of salted water ready for blanching the vegetables.

Scoop out balls of the vegetables using a butter baller or melon scoop. There will be a fair amount of wastage as the balls should be as neat and evenly sized as possible. If preferred, cut the vegetables into small cubes or dice. Save the remaining pieces of vegetable to chop up for soups, etc. Blanch the carrot balls for 5 minutes; the potato ones for 1. Drain well. Fry the shallot gently in the clarified butter for 5 minutes while stirring, until wilted and tender but not coloured. Add the sprig of thyme and the other vegetables; increase the heat to moderately high and sauté for 5 minutes until they are tender but still very firm. Season with salt and pepper. Serve.

DEEP-FRIED POTATOES

1½ lb (700 g) very large waxy potatoes

light olive oil or peanut oil

salt

Peel the potatoes and cut into slices ¼ in (½ cm) thick. Rinse well, drain thoroughly and spread them out on a clean cloth, cover with another cloth and blot completely dry. Heat plenty of oil in a deep-fryer to 300°F (150°C) and carry out the first frying of the potatoes, in batches, leaving them in the oil for 8 minutes or until they bob up to the surface and have just begun to puff up slightly. As each batch is fried, take it out with a slotted spoon or frying basket and spread out in a single layer on kitchen paper. You can complete these first stages an hour or two in advance.

Just before you want to serve the potatoes, heat the oil in the deep-fryer again, this time to 350°F (180°C) and again fry the potatoes in batches until they puff up well and are crisp and golden brown. Drain well, keep hot in a very wide, uncovered heated dish while you fry the rest. Sprinkle lightly with salt and serve at once in a hot, uncovered dish or directly on to heated individual plates.

———•———

POTATO MOUSSELINE

1 ¾ lb (800 g) floury potatoes

approx. 1 pint (500–600 ml) milk

2 oz (50 g) butter

nutmeg

salt and pepper

Serve with:

roast veal or pork, egg dishes or Peas with Bacon and Basil (see page 234)

Peel the potatoes, cut them into fairly small pieces and boil in salted water for 20 minutes or until very tender. Drain well and put twice through the potato ricer or push through a fine sieve. Bring the milk slowly to the boil. Melt the butter in a saucepan, add the potato and a pinch each of salt, pepper and nutmeg. Add the scalding hot milk to the potato in a thin stream, beating continuously with a balloon whisk or hand-held electric beater. Continue beating until the purée reaches boiling point, adding a little more hot milk if necessary. Taste and adjust seasoning. Take off the heat and keep beating for a minute or two longer, by which time the potato should be very light and fluffy. This is a marvellously light, digestible way of preparing potatoes.

———•———

POTATO GRATIN

1¼ lb (600 g) waxy potatoes

1½ oz (40 g) butter

2 sprigs thyme

14 fl oz (400 ml) milk

1 large clove garlic, peeled

2 oz (50 g) grated Gruyère or Emmenthal cheese

2 oz (50 g) grated Parmesan cheese

nutmeg

salt and pepper

Serve with:

egg dishes or roast meat or poultry

Preheat the oven to 400°F (200°C) mark 6. Peel and rinse the potatoes; slice about ¼ in (just under 2 cm) thick using a mandoline cutter if you have one. Heat 1 oz (30 g) of the butter in a wide, non-stick frying pan with the thyme sprigs, and when it is hot add the potatoes and sauté them for 2 minutes, turning with a non-stick spatula. Season with salt, pepper and a pinch of freshly grated nutmeg. Pour in the milk. Heat until the milk is simmering and then continue

197

cooking over a moderate heat for about 10 minutes, stirring frequently with the spatula to prevent the slices catching and burning on the bottom of the frying pan.

Cut the garlic clove lengthwise in half and rub the cut surfaces hard all over the inside of a wide, shallow ovenproof dish. Use some of the remaining butter to grease the inside of this dish lightly.

When the potatoes are cooked and the milk has acquired a thicker, creamy consistency, season with a little more salt and pepper and transfer to the prepared dish. Mix the two cheeses and sprinkle over the potatoes, dot the surface with a few flakes of butter. Place in the oven for 15 minutes, until the cheese topping is crisp and light golden brown. Serve at once. This goes well with roasts, or egg dishes; also delicious when served with green vegetables as a vegetarian main course.

———•———

ALMOND CROQUETTE POTATOES

1 lb (500 g) floury potatoes
2 oz (50 g) butter
3 eggs
6–7 oz (180–200 g) flaked or slivered almonds
nutmeg
light olive oil
3 tbsp finely grated Parmesan cheese
salt and pepper
Serve with:
roast veal, foil-baked trout or Radish and Cress Salad in Cream Dressing (see page 163)

Steam the potatoes, spear and peel them while still boiling hot and put through the potato ricer twice (or push through a fine sieve) while still hot. Transfer to a heavy-bottomed saucepan and keep stirring over a moderate heat as you add the butter, a small piece at a time. Continue stirring until the

mixture is firm and quite dry, leaving the sides of the pan clean, take off the heat and gradually beat in 1 egg yolk and the cheese. Season with salt, freshly ground white pepper and a pinch of nutmeg. Allow to cool completely.

Spread the almonds out in a shallow dish or plate; beat the 2 remaining whole eggs in a deep dish with a pinch of salt.

Shape the potato mixture between your palms into balls the size of walnuts, dip in the beaten egg, drain briefly over the dish and coat evenly with the almonds, pressing down gently so that they adhere. Place on kitchen paper on top of a chopping board or work surface.

Heat plenty of oil in a deep-fryer and lower a few of the croquettes into the oil in the frying basket. Fry for about 3 minutes or until the almond covering is golden brown. Keep hot, uncovered, on kitchen paper while you fry the remaining batches. Serve without delay.

———•———

JERUSALEM ARTICHOKES WITH CREAM SAUCE

1¼ lb (600 g) Jerusalem artichokes
2 shallots, finely chopped
juice of 1 lemon
scant 1 oz (20 g) butter
1 tbsp sunflower oil
1 sprig thyme or small pinch dried thyme
3½ fl oz (100 ml) vegetable stock (see page 38)
5 fl oz (150 ml) crème fraîche or double cream
½ chicken stock cube
2 tbsp chopped chives
salt and pepper

Rinse the Jerusalem artichokes and peel, dropping immediately into a large bowl of cold water acidulated with the lemon juice to prevent discoloration.

198

When they are all peeled, drain them and cut into slices about ¼ in (½ cm) thick while you fry the shallot very gently until wilted and tender in the butter and oil with the thyme. Add the artichokes, stir for 2–3 minutes, add the stock and simmer, uncovered, for 5 minutes. Add the cream, the crumbled ½ chicken stock cube and stir. Simmer uncovered, stirring now and then for a further 5 minutes or until the Jerusalem artichokes are tender but still firm. Season with salt and pepper to taste, sprinkle with the chives and serve at once.

———•———

CARROT CAKE

7 oz (200 g) very finely grated carrots
5 eggs
6 oz (180 g) caster sugar
7 oz (200 g) ground almonds
2 oz (60 g) cornflour or potato flour
1 lemon
butter for greasing
icing sugar
9 fl oz (250 ml) whipping cream
Decorate with:
1 small carrot made out of marzipan or 1 bunch redcurrants

Preheat the oven to 350°F (180°C) mark 4. Beat the egg yolks with half the sugar until pale and creamy. Beat the egg whites until stiff in a separate bowl and gradually beat in the remaining sugar. Mix the almonds with the cornstarch or potato flour and add to the beaten egg yolks together with the carrots and the grated rind and juice of the lemon. Stir very well before folding in the sweetened beaten egg whites. Grease a 10-in (25-cm) diameter spring-release cake tin with butter; transfer the mixture to it, levelling the surface with a palette knife. Bake for 40 minutes. Allow to cool to lukewarm before un-

moulding. Dust with sifted icing sugar and decorate with a marzipan carrot or with a bunch of redcurrants. Serve with the cream, beaten until it just holds it shape.

———•———

INDIAN CARROT PUDDING
Gajjar halva

2¼ lb (1 kg) tender young carrots
1¾ pints (1 litre) full-cream milk
8 whole cardamom pods
4–5 oz (120–150 g) ghee (clarified butter, see page 37)
4–6 tbsp caster sugar
1½ tbsp sultanas
1½ tbsp flaked or slivered almonds
10 fl oz (300 ml) whipping cream, lightly beaten
Decorate with:
few small mint leaves

Trim the carrots and peel with the potato peeler. Grate finely in the food processor or with a grater. Place in a large, heavy-bottomed saucepan or fireproof casserole dish and add the milk. Use the tip of a small knife to make a small slit in the outer covering of the cardamom pods and add to the carrot and milk; they will flavour the mixture subtly without releasing their seeds.

Bring slowly to the boil over a moderate heat; reduce the heat and simmer gently for 30 minutes or until the milk has disappeared, mainly absorbed by the carrots, partly evaporated. Stir in all but 1 tbsp of the ghee and keep stirring while cooking over a very low heat until the mixture turns a russet colour. Stir in sugar to taste; mix well over the heat.

Roast the almonds with the sultanas in a non-stick frying pan without any fat for 1–2 minutes while stirring, then mix into the carrot mixture for 2 minutes. Remove from the heat and allow to cool a little. Serve warm or at room temperature in individual small bowls, with cream beaten until it just holds its shape and decorate each serving with a couple of mint leaves. This is a very filling dessert, best served in small portions. Serves 8.

———— • ————

SWEET POTATO CREAM INDIAN STYLE

9 oz (250 g) sweet potatoes

5 cardamoms

1¾ pints (1 litre) milk

3½ oz (100 g) caster sugar

4 tbsp flaked or slivered almonds, roasted

saffron threads

Decorate and flavour with:

2 tbsp finely chopped, unsalted pistachio nuts

Heat the milk slowly to boiling point in a large, heavy-bottomed saucepan or fireproof casserole dish. Slit open the thick, dry skin of 3 of the cardamom pods, take out all the seeds and grind them finely, using a pestle and mortar or electric grinder. Set aside.

Peel the sweet potatoes, cut into small pieces and process in the food processor until finely grated. Stir into the boiling hot milk and cook over a low heat, stirring frequently, for about 15 minutes. Add the sugar and stir for a further 2 minutes to give the sugar time to dissolve completely. Remove from the heat and leave to cool a little.

Place a generous pinch of saffron threads in a cup, add 2–3 tbsp boiling water and stir with a teaspoon until they dissolve. Stir into the tepid potato mixture, together with the ground cardamom seeds and the almonds.

Spoon into small bowls or coupes, cover with cling film and chill in the refrigerator for 2 hours or more before serving. Just before serving, sprinkle with chopped pistachio nuts.

CARNIVAL FRITTERS

1 large floury potato weighing approx. 9 oz (250 g)

⅔ oz (20 g) fresh baker's yeast or ½ oz (10 g) dried yeast

1 lb 2 oz (500 g) strong flour

1 egg

1 ripe orange

3 tbsp dark Jamaica rum

light olive oil

caster sugar

salt

Peel the potato, cut into fairly small pieces and boil until tender; drain very thoroughly. Put through the potato ricer twice or push through a very fine sieve. Place in a large mixing bowl. Dissolve the yeast in approx. 2 fl oz (50 ml) lukewarm water with ½ tsp caster sugar. Leave to stand for a few minutes if dried yeast is used, until there is a layer of foam on the surface. Sift the flour on to the potato in the bowl with a pinch of salt and stir well, moistening with the lightly beaten egg, the juice and finely grated rind of the orange, the rum and the gently stirred yeast mixture. Add a little more lukewarm water if necessary. The dough should be quite firm and smooth. Shape it into a ball, cover the bowl with a damp cloth and leave to rise in a warm place for about 1 hour or until it has doubled in volume.

Break off pieces of the dough and roll into long, even sausages about ½ in (1 cm) thick; cut these into 10-in (25-cm) lengths, join the ends, pressing them together firmly. Heat plenty of oil in a deep-fryer to 320°F (160°C) and fry the rings, 2 or 3 at a time, until they are golden brown all over. Drain well and keep hot on kitchen paper while frying the rest. Sprinkle liberally with sugar and serve while still very hot.

———— • ————

·FRESH AND DRIED LEGUMES·

PAKORAS

p. 209

Preparation: 1 hour
Cooking time: 20 minutes
Difficulty: Easy
Snack/Appetizer

FRIED STUFFED PURIS

Urd dhal puri

p. 209

Preparation: 45 minutes + 4 hours
soaking time
Cooking time: 25 minutes
Difficulty: fairly easy
Snack/Appetizer

VEGETABLE SAMOSAS

p. 210

Preparation: 1 hour 30 minutes
Cooking time: ½–2 minutes
Difficulty: fairly easy
Snack/Appetizer

BAKED TOMATOES WITH BROAD BEAN MOUSSE

p. 211

Preparation: 30 minutes
Cooking time: 25 minutes
Difficulty: easy
Appetizer

HUMMUS

p. 212

Preparation: 15 minutes + 12 hours
soaking time
Cooking time: 3 hours
Difficulty: very easy
Appetizer

BEAN AND RADICCHIO SALAD WITH HORSERADISH DRESSING

p. 212

Preparation: 10 minutes
Difficulty: very easy
Appetizer or accompaniment

SPLIT GREEN PEA SOUP

p. 213

Preparation: 20 minutes + 2 hours 30
minutes soaking time
Cooking time: 1 hour
Difficulty: easy
First course

INDIAN PEA SOUP

Hara shorba

p. 213

Preparation: 25 minutes
Cooking time: 45 minutes
Difficulty: very easy
First course

PEA AND ASPARAGUS SOUP

p. 214

Preparation: 15 minutes
Cooking time: 20 minutes
Difficulty: very easy
First course

VENETIAN RICE AND PEA SOUP

Risi e bisi

p. 214

Preparation: 25 minutes
Cooking time: 30 minutes
Difficulty: easy
First course

HARICOT BEAN SOUP

p. 215

Preparation: 15 minutes + 12 hours
soaking time
Cooking time: 1 hour 30 minutes
Difficulty: easy
First course

BUTTER BEAN AND PUMPKIN SOUP

p. 216

Preparation: 20 minutes
Cooking time: 1 hour
Difficulty: easy
First course

Bean and radicchio salad with horseradish dressing

Springtime risotto

BEANS AND EGGS MEXICAN STYLE

Frijoles con huevos rancheros a la mexicana

p. 226

Preparation: 30 minutes
Cooking time: 3 hours
Difficulty: very easy
Main course

WARM BEAN SALAD

p. 226

Preparation: 10 minutes
Cooking time: 2 hours
Difficulty: very easy
Main course

FRENCH BEAN AND POTATO PIE

p. 227

Preparation: 30 minutes
Cooking time: 45 minutes
Difficulty: easy
Main course

CANNELLINI BEANS WITH HAM AND TOMATO SAUCE

p. 228

Preparation: 20 minutes
Cooking time: 2 hours 30 minutes
Difficulty: easy
Main course

Baked tomatoes with broad bean mousse

NORTH AFRICAN VEGETABLE CASSEROLE

p. 228

Preparation: 45 minutes + 12 hours soaking time
Cooking time: 3 hours 30 minutes
Difficulty: easy
Main course

GREEK CHICK PEA CROQUETTES

Pittarúdia

p. 229

Preparation: 20 minutes + 12 hours soaking time
Cooking time: 25 minutes
Difficulty: easy
Main course

MUNG DHAL FRITTERS

p. 229

Preparation: 30 minutes + 12 hours soaking time
Cooking time: 25 minutes
Difficulty: easy
Main course

BORLOTTI BEAN PURÉE

p. 231

Preparation: 10 minutes
Cooking time: 2 hours 30 minutes
Difficulty: easy
Main course or accompaniment

MEXICAN BEANS WITH GARLIC AND CORIANDER

p. 231

Preparation: 15 minutes
Cooking time: 2 hours 30 minutes
Difficulty: very easy
Main course or accompaniment

BROAD BEANS WITH CREAM

p. 232

Preparation: 20 minutes
Cooking time: 20 minutes
Difficulty: very easy
Main course or accompaniment

JAPANESE STIR-FRIED RICE WITH GREEN PEAS AND MIXED VEGETABLES

Yakimeshi

p. 232

Preparation: 30 minutes
Cooking time: 6 minutes (with ready-cooked rice)
Difficulty: easy
Accompaniment or appetizer

Warm bean salad

Split green pea soup

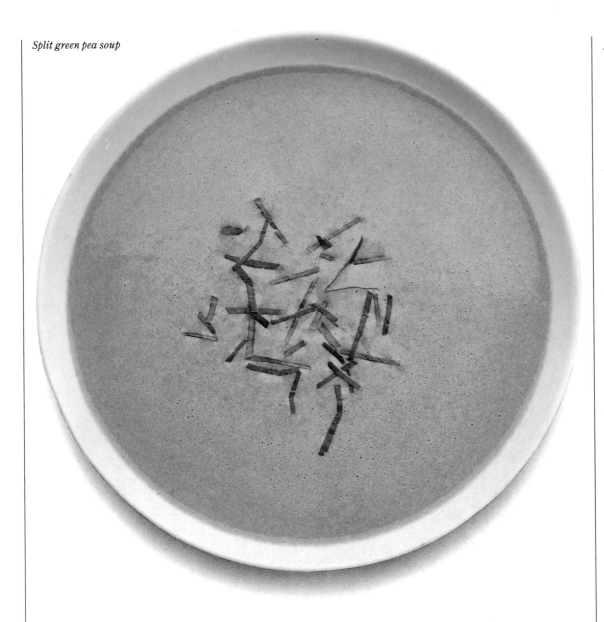

PAKORAS

For the chick pea flour batter:

9 oz (250 g) besan (chick pea) flour

scant ½ tsp baking powder

½ tsp cayenne pepper

½ tsp ground coriander seeds

1 tsp turmeric

salt

For the coconut chutney:

approx. 8 oz (225 g) grated fresh coconut flesh

½ clove garlic

2 tbsp grated fresh ginger

1 green chilli pepper, finely chopped

6 tbsp chopped coriander leaves

2 fl oz (60 ml) Greek yoghurt

scant 1 tbsp fresh lime juice

½ tsp black mustard seeds

1 tbsp sunflower or peanut oil

salt

For the spicy ketchup:

7 fl oz (200 ml) tomato ketchup

½–1 tsp pili-pili hot relish (see page 113)

For the fritters:

1 large onion

1 sweet potato

2 oz (60 g) spinach leaves (trimmed net weight)

1 medium-sized aubergine

12 okra or 3 green peppers

sunflower oil

Ask for besan flour at your nearest Indian grocer's shop. It is pale, creamy yellow in colour. Make the batter. Sift the flour, baking powder, cayenne pepper, coriander, turmeric and a pinch of salt together into a large mixing bowl. Gradually add just enough cold water to make a fairly thick coating batter, stirring continuously with a wooden spoon. Add a little more salt to taste, cover and leave to stand.

Make the coconut chutney: place the grated coconut flesh in the blender with 2 fl oz (60 ml) cold water, the garlic, ginger, chilli pepper and coriander and process at high speed until smooth, adding a little more water if necessary. Spoon into a serving bowl and mix in the yoghurt, lime juice and a pinch of salt. Heat the oil in a non-stick frying pan and fry the mustard seeds for a few seconds, stirring them as they cook; as soon as they start to jump about in the pan, remove from the heat and stir them into the coconut mixture; add a little more salt if needed and place the bowl in the refrigerator to chill until just before serving. Mix the ketchup with the pili-pili relish, spoon into 4 very small bowls and set aside.

Prepare the vegetables. Peel the onion and the sweet potato; wash the spinach leaves and drain well. Cut the onion into thin rings and the aubergine and sweet potato into thin round slices. Heat plenty of sunflower oil in a deep-fryer to 350°F (180°C) or use a wok. Dip the vegetable pieces one by one in the batter and then add to the very hot oil. Fry in small batches until the batter coating has puffed up and is crisp and golden brown. Remove from the deep-fryer or wok and keep hot, uncovered, on kitchen paper to finish draining while you fry the remaining batches. Serve at once with the coconut chutney and spicy ketchup.

———•———

FRIED STUFFED PURIS
Urd dhal puri

6 oz (180 g) split black gram (urd dhal, see method)

4½ oz (130 g) chapati flour (see method) or plain flour

5 tbsp ghee (clarified butter, see page 36)

1 green chilli pepper, finely chopped

sunflower oil

salt

For the garam masala:

10 black peppercorns

2 tsp coriander seeds
2 tsp fennel seeds
2 tsp cumin seeds
Serve with:
7 fl oz (200 ml) natural yoghurt, lightly salted
1 spring onion, sliced

The above quantities make 5 stuffed puris. You can buy black gram (urd dhal) from Indian grocers and from health food stores which stock a good range of pulses. Check that the split gram is white, not yellow. Rinse in a sieve under running cold water; drain and then soak in a bowl of cold water for 4 hours. While they are soaking, make the dough.

If possible, use chapati (ata) flour, a wholemeal flour sold by Indian grocers and by many specialist food stores. Sift it into a large mixing bowl, add ½ tsp salt, 2 tbsp hot ghee and stir in, gradually adding just enough cold water to make a firm dough. Knead for 10 minutes then shape into a ball, wrap in cling film and leave to rest for at least 30 minutes.

Make the garam masala: heat a non-stick frying pan, add the peppercorns and the coriander, fennel and cumin seeds and toast for 30 seconds over a moderate heat, stirring continuously. Remove from the heat; grind the seeds finely in a spice mill or use a pestle and mortar; put through a fine sieve to trap the pieces of coriander skins and discard these.

Once the lentils have finished soaking, place them in the food processor with a little cold water and process briefly until they form a coarse, thick purée. Heat the remaining ghee in a large frying pan and fry the chopped chilli pepper gently over a low heat for 2–3 minutes, stirring continuously; add the lentil purée and ½ tsp salt. Cook, stirring over a slightly higher heat for a few minutes, until the purée darkens and thickens further. Remove from the heat and leave to cool before stirring in the garam masala.

Start making the puris about 30 minutes before you plan to serve them. Knead the dough for 2–3 minutes and divide into 10 pieces of even size; shape these into balls between your palms and drop them into a large bowl; remove 2 and cover the bowl with

cling film to prevent the dough from drying out. Flatten these first 2 balls with your palms and then roll out on a lightly floured pastry board with a floured rolling pin to form 2 thin, matching discs just over 3 in (8 cm) in diameter. Have a cup of cold water standing ready. Place 2 tbsp of the lentil filling in the centre of one disc, dipping your finger into the cold water, moisten all round the edge of the disc and then cover neatly with the other disc, pressing the edges together to achieve a good seal. Do exactly the same with the remaining dough and filling. You will have 5 puris ready for deep-frying.

Heat plenty of oil in a very large frying pan or deep-fryer and when hot but not smoking, place 3 of the puris in the pan and fry until they puff up and are pale golden brown; keep spooning the very hot oil over the exposed surfaces of the puris. Drain and keep hot while you fry the remaining 2. Serve at once, with a bowl of *raita* made by mixing the yoghurt with a pinch of salt and with the finely sliced spring onion.

———•———

VEGETABLE SAMOSAS

For the pastry dough (makes 16 samosas):
9 oz (250 g) plain flour
6 tbsp ghee (clarified butter, see page 36)
2 fl oz (50 ml) hot milk
sunflower oil
salt
For the filling:
¾ lb (350 g) frozen peas, thawed
14 oz (400 g) waxy potatoes, boiled
7 oz (200 g) chopped onion
1 small piece fresh ginger, grated
1 green chilli pepper, finely chopped
2½–3 tbsp chopped coriander leaves
6 tbsp ghee (clarified butter, see page 36)

1 tsp coriander seeds, ground and sifted
1 tsp mild or medium curry powder
1 tsp ground toasted cumin seeds
pinch chilli powder
2–3 tbsp fresh lime juice
salt
For the coriander chutney:
3 oz (80 g) fresh coriander leaves
1 green chilli pepper, finely chopped
2 tbsp fresh lime juice
½ tsp ground cumin seeds
salt and pepper

Make the pastry: sift the flour into a large mixing bowl, add 2 tsp salt, 5 tbsp of the ghee and rub in with your fingertips until the mixture resembles breadcrumbs. Gradually work in as much of the hot milk as is needed to make a firm dough. Knead for a few minutes until the dough is smooth and elastic. Shape into a ball, brush the surface lightly with oil all over and wrap in cling film to prevent it drying out. Leave to rest for 45 minutes.

While the pastry dough is resting, make the filling. When the potatoes have cooled completely peel and dice. Heat 4–6 tbsp of the ghee in a non-stick frying pan and sweat the onion for 15 minutes, stirring frequently. When it is tender but not browned, add the peas, grated ginger, chilli pepper, chopped coriander leaves and 3½ fl oz (100 ml) water. Stir, adding a pinch of salt, cover and simmer gently for about 20 minutes or until the peas are very tender. Add the diced potato, the ground coriander and cumin, the curry powder, chilli powder and fresh lime juice to taste. Stir briefly and then remove from the heat.

Make the coriander chutney. Wash and drain the coriander sprigs and detach the leaves together with their smallest, tenderest stalks; measure out the required amount and place in the food processor. Remove the stalk, seeds and inner pale parts of the chilli pepper, chop and add to the process with the lime juice, scant 1 tsp salt, ground cumin and

some freshly ground black pepper; add scant 1 tbsp cold water and process at high speed to form a thick paste. Spoon into a small bowl.

Assemble the samosas: divide the pastry dough into 8 equal parts and roll into balls between your palms. Take up one of them and place the others in a large bowl, covered with cling film to prevent them drying out. Flatten the ball of dough between your palms and then roll out on a lightly floured pastry board with a floured rolling pin to make a thin disc of pastry about 7 in (18 cm) in diameter. Cut this disc in half, moisten along the cut edge and fold the semi-circle in half, pressing the moistened edges together: you will now have a triangular pocket. Hold this pastry case in one hand and use a teaspoon to fill it, packing in the filling gently; when it is full, moisten along the inside top edges and pinch them together to seal. Place the samosa flat on the pastry board and press along the sealed edges with the prongs of a fork to complete the sealing process and give the edges a decorative, ribbed effect. Place on a chopping board and cover with cling film to prevent the pastry drying out while you prepare a further 15 samosas in the same way.

Heat plenty of sunflower oil in a deep fryer to 350°F (180°C) or use a wok or deep, heavy-bottomed saucepan. Fry the samosas in batches until golden brown (about 1½–2 minutes). Drain and serve the coriander chutney.

———— • ————

BAKED TOMATOES WITH BROAD BEAN MOUSSE

1 lb (500 g) young, tender broad beans (net shelled weight)
2 lb (900 g) medium-sized ripe tomatoes
1 egg
2 oz (60 g) freshly grated Parmesan cheese
½ clove garlic, peeled and finely grated
1 tinned anchovy fillet

2 tbsp extra-virgin olive oil

fine breadcrumbs

salt and pepper

Bring a saucepan of lightly salted water to the boil ready to cook the broad beans. Preheat the oven to 400°F (200°C) mark 6. Rinse and dry the tomatoes; slice off and discard a 'lid' from each one.

Remove the seeds and the dividing sections of the flesh as well as the central part. Lightly oil a baking tray or shallow ovenproof dish and place the tomatoes in it. Sprinkle some salt inside each tomato.

When the water in the saucepan comes to the boil, add the broad beans and cook them until very tender; drain them and remove their seed cases when cool enough to handle. This will take some time; do not worry if you break up their contents. Discard the seed cases. Place the broad beans in the food processor, add the lightly beaten egg, a little salt and pepper, the Parmesan cheese and the garlic. Process to a smooth purée. Transfer to a bowl.

Crush the anchovy fillet into 2 tbsp oil over a low heat, using the back of a wooden spoon; when it is mashed, stir well into the broad bean mixture. Add sufficient fine soft breadcrumbs to thicken the broad bean mousse further: it should hold its shape firmly. Spoon it in batches into a forcing bag fitted with a medium-sized, fluted nozzle. Pipe mousse into each tomato (see illustration on page 207). Bake for 7–10 minutes.

———•———

HUMMUS

1 lb (450 g) chick peas

8–10 tbsp tahini (sesame seed paste)

4 cloves garlic, peeled and finely grated

4–5 fl oz (120–150 ml) fresh lemon juice

cayenne pepper

1 sprig parsley

extra-virgin olive oil

salt

Serve with:

hot pitta bread

Rinse the chick peas thoroughly in a sieve under running cold water, drain, place them in a large bowl and add sufficient cold water to cover them amply. Leave to soak for at least 12 hours. When they have finished soaking, drain them, place in a large, heavy-bottomed saucepan or flameproof casserole dish, pour in enough cold water to come about 2 in (5 cm) above them and bring slowly to the boil. Turn down the heat to very low, place a lid on the saucepan, leaving a small chink for the steam to escape and simmer gently for 3 hours. Do not add salt until they have cooked for approx. 2¾ hours. Drain, keeping their cooking liquid for soups, etc. as it is very nutritious. Reserve about 6 whole chick peas with which to decorate the finished hummus; put all the rest through the vegetable mill, with a fine-gauge disc fitted, into a large mixing bowl.

Beat in the tahini, the garlic and lemon juice to taste. Beat in a little of the reserved cooking liquid. Add salt to taste and transfer the hummus to a serving bowl.

Sprinkle the surface with a pinch of cayenne pepper, arrange the reserved whole chick peas in a circle in the centre around the sprig of parsley and drizzle a little olive oil over the surface.

———•———

BEAN AND RADICCHIO SALAD WITH HORSERADISH DRESSING

1 lb (500 g) cooked fresh or tinned borlotti beans

10 oz (300 g) radicchio

1 tbsp grated horseradish mixed with 1 tbsp wine vinegar or cider vinegar

5–6 tbsp olive oil

salt and freshly ground white or black pepper

Serve with:

cold meat

Prepare the radicchio as described on page 22. Separate the leaves and use as containers for the beans or cut the radicchio heads and bases into fairly small pieces with a very sharp stainless steel knife; rinse and drain well. Drain the beans and mix with the radicchio in a large salad bowl or arrange as shown in the illustration on page 201.

Mix the horseradish and vinegar with ½ tsp salt and plenty of freshly ground pepper, then gradually beat in the olive oil, adding a very little at a time, so that it emulsifies well. Mix this dressing with the salad or sprinkle over it and serve.

———— • ————

SPLIT GREEN PEA SOUP

1 lb (450 g) dried split green peas

1 small onion

1 medium-sized carrot

1 green celery stalk

1 leek

2 oz (60 g) lean streaky bacon rashers

1 bouquet garni

2 pints (1.2 litres) chicken stock (see page 37)

1 oz (25 g) fresh chervil or watercress leaves

3½ fl oz (100 ml) crème fraîche or double cream

1 oz (30 g) unsalted butter

2 tbsp sunflower oil

salt and pepper

Soak the peas in a large bowl of cold water for 2½ hours, drain well in a sieve, rinse well under running cold water, drain again and place in a large saucepan with 1¼ pints (700 ml) water ready to heat to boiling. While the peas are soaking, coarsely chop the peeled onion and carrot together with the celery, leek and bacon. Sweat all these chopped ingredients in 2 tbsp oil for 15 minutes over a gentle heat, stirring frequently. Add this mixture to the pre-soaked split peas and water in the saucepan together with the bouquet garni. Bring to the boil, then partially cover so as to allow the steam to escape, reduce the heat to low and simmer for 40 minutes, by which time the peas should be very tender. Discard the bouquet garni; beat the soup with a hand-held electric beater until it is smooth, or put through a vegetable mill and return to the saucepan. Add the hot stock to this purée until the consistency you prefer is achieved. Season to taste and bring slowly to the boil.

Rinse the chervil or watercress; blanch for a few seconds only in boiling salted water, drain, refresh under running cold water, squeeze out excess moisture and reserve. Stir the cream into the soup and remove from the heat as soon as it returns to a gentle boil. Add a little more salt if wished, stir in the chervil or watercress and the butter.

———— • ————

INDIAN PEA SOUP
Hara shorba

10 oz (300 g) frozen peas

2 potatoes, peeled and diced

1 onion, peeled and diced

2 oz (60 g) spinach leaves (net trimmed weight)

2½ tsp cumin seeds

½ tsp coriander seeds

1 piece fresh ginger, 1¼ in (3 cm) thick

2¼ pints (1.3 litres) chicken stock (see page 37)

2 tbsp coriander leaves

½ green chilli pepper, finely chopped

2 tbsp fresh lime juice

6 mint leaves

5 fl oz (150 ml) single cream

pinch ground cinnamon

salt and pepper

Garnish with:

coriander leaves

Clean and prepare all the vegetables. Grind the cumin seeds in a spice mill, or pound to a powder with a pestle and mortar. Take out ½ tsp of the ground cumin and set aside; add the coriander seeds to the cumin in the mill or mortar and grind them too. Sift in a fine sieve to separate the coriander seed cases. Place ½ tsp ground cumin in a small, non-stick frying pan and toast gently over a low heat, stirring; as soon as the spice releases a full aroma, remove from the heat and reserve.

Place the potatoes, onion, the whole piece of ginger and the mixed cumin and coriander in a large saucepan with the chicken stock. Bring to the boil, cover and simmer over a fairly low heat for 20 minutes. Add the peas, the well washed spinach, fresh coriander leaves, mint, the reserved toasted cumin, the chilli pepper and 1 tsp salt. Replace the lid and simmer for a further 25 minutes. Remove and discard the ginger; beat the soup with an electric beater to reduce to a purée or put through a vegetable mill and return to the saucepan. Stir in the cream, add a little more salt if needed, add lime juice to taste, a pinch of cinnamon and some freshly ground pepper. Garnish with the coriander leaves.

———•———

PEA AND ASPARAGUS SOUP

7 oz (200 g) peas, fresh or frozen

5 oz (150 g) asparagus tips

2 spring onions

2 medium-sized new potatoes

1 small lettuce

2 pints (1.2 litres) vegetable stock (see page 38) or chicken stock (see page 37)

7 oz (200 g) long-grain rice

2 tbsp chopped parsley

1 tbsp sunflower oil

generous 1 oz (35 g) unsalted butter

salt and pepper

Bring the stock to the boil. Prepare the vegetables: use the tender, top half of the well washed asparagus spears; measure out the above net weight and cut them into pieces about ¾ in (2 cm) long. Slice the spring onions. Peel the potatoes and dice. Shred the lettuce into a chiffonade (see page 24).

Heat scant 1 oz (20 g) of the butter with 1 tbsp oil in a heavy-bottomed saucepan and sweat the spring onions until tender. Add the peas and lettuce and cook, stirring, for 1 minute. Pour in the boiling hot stock, stir and allow to return to the boil before sprinkling in the rice. Add the potatoes and asparagus. Stir briefly, cover but leave a gap for the steam to escape, reduce the heat to moderately low and simmer for 14–15 minutes, or until the rice is tender but still fairly firm. Remove from the heat, add the solid remaining butter and the chopped parsley. Stir until the butter has melted and blended. Season and serve.

———•———

VENETIAN RICE AND PEA SOUP
Risi e bisi

10 oz (300 g) peas, fresh or frozen

2 oz (50 g) streaky bacon, finely chopped

1 spring onion, peeled and sliced into rings

3 tbsp finely chopped parsley

2 pints (1.2 litres) boiling hot vegetable stock (see page 38) or chicken stock (see page 37)

7 oz (200 g) long-grain rice

2 oz (60 g) butter

2 tbsp extra-virgin olive oil

6 tbsp freshly grated Parmesan cheese

salt and pepper

This dish is half way between a soup and a risotto. Bring the stock slowly to the boil. While it is heating fry the bacon, spring onion and the parsley in 1 oz (30 g) of the butter and the olive oil in a large, heavy-bottomed saucepan or deep frying pan over a low heat, stirring frequently; when the onion is just tender but not at all browned, add the peas, turn up the heat slightly and cook for 1 minute while stirring.

Pour in about 8 fl oz (450 ml) of the boiling stock, turn down the heat to low, cover and simmer gently for 10 minutes. Add the remaining stock and when it comes to the boil, sprinkle in the rice; cook, uncovered, over a fairly high heat for about 14 minutes or until the rice is tender but still has a little 'bite' left to it, stirring now and then. There should be plenty of slightly thickened liquid left in the pan when the rice is done.

Remove from the heat and stir in the remaining, solid, butter and the Parmesan cheese. Stir gently but thoroughly. Add salt and pepper to taste, stir once more, cover and leave to stand for 5 minutes. Serve in heated individual soup bowls or deep plates.

—— • ——

HARICOT BEAN SOUP

1 lb (500 g) dried haricot beans, soaked

1½ tbsp extra-virgin olive oil

½ tsp bicarbonate of soda

salt

For the sauce:

1 large clove garlic, peeled and slightly crushed

8 medium-sized ripe tomatoes, peeled and seeded

½–1 dried chilli pepper (optional)

pinch oregano

2 tbsp extra-virgin olive oil

salt and freshly ground pepper

Garnish with:

4 sprigs basil

Serve with:

garlic bread

Start preparation the day before you plan to serve this dish. Rinse the beans in a sieve and place in a bowl with enough cold water to amply cover them; add ½ tsp bicarbonate of soda. Leave the beans to soak overnight or for more than 12 hours, then drain, place in a saucepan with enough water to cover them completely and bring slowly to the boil over a moderate heat. Cover, leaving a gap so that the steam can escape. Simmer very gently for about 1½ hours or until the beans are very tender. Remove the scum from the surface of the water at intervals. Add a large pinch of salt when the beans are nearly done.

While the beans are cooking, make the sauce: heat the olive oil in a saucepan over a moderately low heat and add the whole garlic clove, partially crushed with a flat of a heavy knife blade. After about 1 minute, when the garlic is pale golden brown, add the chopped tomatoes. Turn up the heat, add the seeded, crumbled chilli pepper and a pinch of oregano; cook briskly, stirring, for 2–3 minutes. Set aside.

When the beans are done, put them with their cooking liquid through the vegetable mill with a fine-gauge disc fitted. Ladle the resulting purée into individual heated bowls, spoon some of the tomato mixture into the centre of each bowl, drizzle a little olive oil on to each serving and sprinkle with a generous amount of freshly ground black pepper. Garnish with the basil sprigs and serve with garlic bread.

—— • ——

Butter Bean and Pumpkin Soup

3½ oz (100 g) dried butter beans

1 lb (500 g) pumpkin flesh (net weight)

2 large floury potatoes

1¾ pints (1 litre) milk

3½ oz (100 g) long-grain rice

1 vegetable or chicken stock cube

½ tbsp caster sugar

2 tbsp chopped parsley

salt and white peppercorns

Rinse the beans and soak them in a bowl of cold water overnight or for at least 12 hours. Drain them and place in a saucepan with plenty of cold water; bring slowly to the boil then simmer gently for 45 minutes, adding salt shortly before the end of this cooking time. (If you prefer to use tinned beans, drain and use approx. 8 oz (225 g).

While the beans are cooking, peel the pumpkin and remove the seeds and filaments from the pumpkin; weigh the flesh and cut into small pieces. Cut the peeled potatoes into pieces of the same size. Place the pumpkin and potato pieces in a saucepan, add 1½ pints (900 ml) water and 1 tsp salt. Bring to the boil, cover and simmer gently for 40 minutes. Take the pan off the heat and use a hand-held electric beater to reduce to a purée, or put through a food mill and return to the saucepan.

Stir in the milk and beans and place on a moderate heat. When the liquid comes to the boil, add the rice and the stock cube, stir well, turn down the heat and simmer, uncovered, for 12–13 minutes, stirring at frequent intervals. Test the rice; it should be tender but still firm. Stir in the sugar, a little salt if needed and add plenty of freshly ground white pepper.

Ladle into soup bowls and sprinkle with the parsley.

—— • ——

Minestrone

1 lb (500 g) peas, fresh or frozen

8 oz (250 g) fresh or tinned borlotti beans

8 oz (250 g) ripe tomatoes, skinned, seeded and chopped

3 large potatoes, peeled and left whole

2 young courgettes, sliced into rounds

2 stalks green celery, sliced thinly across the stalks

2 medium-sized carrots, sliced into rounds

¼ Savoy cabbage, cut into large pieces

2 oz (50 g) salt pork back fat, finely chopped

1 clove garlic, peeled and finely chopped

1 medium-sized onion, peeled and finely chopped

3 tbsp chopped parsley

2 sage leaves, finely chopped

4 oz (120 g) streaky bacon rashers, chopped

10 basil leaves, finely chopped

7 oz (200 g) risotto rice (e.g. arborio)

approx. 3 oz (80 g) grated Parmesan cheese

extra-virgin olive oil

salt and pepper

Prepare all the vegetables. Chop the pork fat, garlic and parsley together. Cut the bacon into strips. Gently fry the chopped fat, garlic and parsley in a very large, heavy-bottomed saucepan together with the onion, sage and the bacon, stirring frequently. When these have cooked for 15–20 minutes and the onion is soft and very lightly browned, add all the other vegetables, including the beans and the basil, but reserving the peas and the cabbage which will be added later. Pour in 5½ pints (3 litres) cold water, add 2 tsp sea salt and bring slowly to the boil. Cover, reduce the heat to very low and simmer gently for 1½ hours, skimming off any scum that rises to the surface. Use a slotted spoon to take the potatoes out of the soup, mash them coarsely in a bowl with a fork or potato masher and return to the saucepan. Add the peas and the cabbage and continue cooking for a further 30 minutes. The soup can be prepared 24 hours in advance up to this point, left to cool and then refrigerated until shortly before serving.

Bring the soup up to a brisk boil, uncovered; sprinkle in the rice and stir. Cook, uncovered over a fairly high heat for 13–14 minutes or until the rice is just tender. Add a little more salt if needed. Hand round a bowl of grated Parmesan cheese for each person to sprinkle into the soup. Serves 6.

———•———

MISO AND TOFU SOUP
Tofu no misoshiru

4 tbsp shiro miso (fermented white soybean paste)
1–1½ tbsp dashinomoto (instant fish stock)
pinch ajinomoto (Japanese taste powder) or monosodium glutamate (optional)
1 cake fresh tofu
½ leek, white part only
Serve with:
steamed rice

Make the stock in a non-metallic saucepan (an enamelled flameproof casserole dish is best) using 1¾ pints (1 litre) water, the instant *dashi* and a pinch of *ajinomoto*; if you do not want to use this taste enhancer, leave it out. Bring to the boil, then reduce the heat to very low. Mix the miso in a bowl with about 4 fl oz (120 ml) of the hot broth; pour the resulting mixture back into the saucepan or casserole dish through a sieve, rubbing with the back of a wooden spoon. Stir for 2 minutes; the soup must not come to the boil once the miso has been added.

Cut the tofu into small cubes, add to the soup and leave to heat through for 2 minutes; remove from the heat. Ladle the soup into bowls, garnish with the leek, cut into strips and serve at once, with plain steamed rice. This makes a very light but nourishing supper.

CHICK PEA AND SPINACH SOUP

1 lb (500 g) chick peas
4½ pints (2½ litres) vegetable stock (see page 38) or chicken stock (see page 37)
8 oz (250 g) spinach
1 small onion
1 clove garlic
1 small carrot
1 celery stalk
2 tbsp plain flour
4 tbsp olive oil
1 tbsp bicarbonate of soda
salt and black peppercorns

Start preparation a day in advance by soaking the chick peas in plenty of cold water to which the bicarbonate of soda has been added. The next day, drain them in a sieve and rinse well under running cold water; place in a saucepan with the stock. Cover, bring slowly to the boil and then simmer gently for 1½ hours.

When the beans have been cooking for about 1¼ hours, wash the spinach, drain and cut into thin strips. Chop the peeled onion, garlic and carrot finely with the trimmed and washed celery and fry gently in 3 tbsp olive oil in a large, heavy-bottomed saucepan or flameproof casserole dish for 5 minutes, stirring with a wooden spoon. Add the flour and stir continuously for 1 minute as it cooks. Add the shredded spinach, stir, draw aside from the heat and add about 8 fl oz (225 ml) of the cooking liquid from the chick peas, beating continuously to prevent lumps forming as the flour thickens the liquid. Add the contents of this saucepan to the pan containing the cooked chick peas. Stir, cover and simmer very gently for a further hour over a very low heat.

Remove three ladlefuls (approx. 12 oz (350 g)) of the beans from the saucepan and reserve. Put the remaining contents of the pan, liquid and beans, through the vegetable mill and return to the pan. Add the reserved, whole beans, season and serve.

217

BEAN AND PUMPKIN RISOTTO

2¼ lb (500 g) dried large white kidney beans or butter beans

2 lb (900 g) pumpkin flesh (net peeled trimmed weight)

2 vegetable or chicken stock cubes

1 small onion, peeled and finely chopped

2½ oz (70 g) butter

1 lb (500 g) long-grain rice

3½ fl oz (100 ml) dry vermouth or dry white white

5 fl oz (150 ml) milk

2½ oz (70 g) freshly grated Parmesan cheese

1½ tbsp finely chopped parsley

salt and pepper

Soak the beans overnight or for about 12 hours in plenty of cold water. Drain in a sieve, rinse well under running cold water and remove their thin seed cases, discarding all these skins. There is no really quick way to do this, but spreading them out between 2 clean cloths and rubbing to break and loosen the skins will help. Place the beans in a very large, heavy-bottomed saucepan with 4½ pints (2½ litres) water and bring slowly to the boil; cover, leaving a gap for the steam to escape, and simmer gently for 1 hour.

Dice the pumpkin flesh. When the beans have been simmering for 1 hour, add the diced pumpkin and simmer for a further 1 hour, by which time the beans should be very tender. Use a hand-held electric beater to turn the contents of the saucepan into a very thin purée, or put through a vegetable mill with a fine-gauge disc fitted. Add the stock cubes, stir well and bring to a gentle boil over a low heat while stirring with a wooden spoon. Set aside.

Fry the onion very gently in 1 oz (30 g) of the butter for 10 minutes, stirring now and then; add the rice, cook while stirring for 1½ minutes over a slightly higher heat, pour in the vermouth or wine and continue cooking until it has completely evaporated. Add about 8 fl oz (225 ml) of the pumpkin and bean purée and continue cooking, uncovered, stirring occasionally and adding more of the purée as the rice

cooks and absorbs the moisture. After 10 minutes, add the hot milk and stir. Cook for a further 4 minutes, or until the rice is tender but still firm; this risotto should be very moist. Remove from the heat, add salt to taste, the remaining, solid butter and the Parmesan cheese and stir gently into the risotto. Season with freshly ground pepper, sprinkle with the parsley and serve. Any remaining purée can be used in vegetable soups. Serves 6.

———— • ————

SPRINGTIME RISOTTO

3½ oz (100 g) peas, fresh or frozen

2 oz (50 g) French beans (net trimmed weight)

3½ oz (100 g) fresh green or white asparagus tips

2 tender baby globe artichokes or 4 fresh artichoke bottoms

1 medium-sized waxy potato

1 medium-sized carrot

3 ripe tomatoes

1 lettuce

1 onion

1 green celery stalk

14 oz (400 g) risotto rice

2½ fl oz (70 ml) dry vermouth or dry white wine

2¾ pints (1½ litres) vegetable stock (see page 38) or chicken stock (see page 37)

2 oz (50 g) freshly grated Parmesan cheese

2 oz (60 g) unsalted butter

4 tbsp olive oil

salt and pepper

Set the stock to come slowly to the boil while you trim, wash and peel the vegetables where necessary. Cut the French beans into pieces just under ½ in (1 cm) long, the asparagus tips into ¾ in (2-cm) lengths. Cut the artichoke hearts if very tender varieties or buds are unavailable use artichoke bottoms) into small cubes; do likewise with the potato, carrot and the blanched, peeled and seeded tomatoes. Cut the lettuce heart into a chiffonade (see page 24); chop the onion and celery together finely. Heat 4 tbsp of the oil and 1 oz (30 g) of the butter in a wide, heavy-bottomed saucepan or very large frying pan and gently fry the chopped onion and celery over a low heat, stirring frequently. Add all the other vegetables except the peas and the asparagus. Cook gently, stirring at frequent intervals, for 10 minutes, moistening with a couple of tablespoonfuls of stock whenever the mixture starts to look a little dry. Stir in the peas and the asparagus; sprinkle in the rice. Turn up the heat to moderately high and cook while stirring to let the rice absorb the fat and flavours in the pan. Add the vermouth and cook until it has evaporated. Pour in about 8 fl oz (225 ml) of boiling hot stock and stir. Continue cooking, uncovered, for about 14 minutes, adding more hot stock as it is absorbed by the rice and reduces. Stir at intervals. Taste the rice to see whether it is done: it should be tender but still slightly firm to the bite. The finished risotto should have plenty of slightly thickened liquid in the pan.

Draw aside from the heat and stir in the remaining, solid butter and the Parmesan cheese. Add salt and pepper to taste, stir gently once more and serve at once. Serves 6.

———•———

PASTA WITH PEAS AND PARMA HAM

10 oz (300 g) peas, fresh or frozen
1 small onion, peeled and finely chopped
1 small clove garlic, peeled and finely chopped
2 thin slices Parma ham, finely chopped
5 basil leaves, finely chopped
3 tbsp olive oil
14 fl oz (400 ml) beef stock (see page 37)

8 oz (250 g) medium-sized short pasta
2 oz (50 g) grated Parmesan cheese
salt and black peppercorns

Set a large pan of salted water to come to the boil. While it is heating chop the onion finely with the garlic, Parma ham and basil. Fry these together very gently in the oil in a large, heavy-bottomed saucepan or flameproof casserole dish for 15 minutes, stirring occasionally. Add the peas and cook for 5 minutes, still over a low heat. Pour in the stock, add a pinch of salt, cover as soon as the liquid has come to the boil and simmer for about 25 minutes, stirring at intervals, by which time the peas will be cooked and very tender.

While the peas are cooking, add the pasta to the boiling salted water and cook until it is only just tender; drain and add to the peas together with some of its cooking water. The finished dish should be very moist. Add a little more salt if necessary, and plenty of freshly ground pepper. Remove from the heat, stir in the Parmesan cheese and leave to stand for 5 minutes before serving.

———•———

PASTA AND BEANS WITH ROCKET SALAD

2¼ lb (1 kg) fresh or tinned borlotti beans
1 small green celery stalk
1 small onion, peeled and quartered
2 cloves garlic, peeled
2 stock cubes
5 oz (150 g) thin egg noodles, broken into short lengths
extra-virgin olive oil
salt and black peppercorns
For the salad:
approx. 8 oz (225 g) young, tender salad rocket

extra-virgin olive oil
wine vinegar
pili-pili hot relish (see page 113) (optional)
salt and pepper
Serve with:
granary bread

If you are using fresh beans, place in a large saucepan and add sufficient water to come 2 in (5 cm) above them. Bring to the boil over a moderate heat and then simmer, uncovered, for 15 minutes, then drain. Return the beans to the saucepan, add 3½ pints (2 litres) water, the celery, onion and garlic and reheat to boiling.

Turn down the heat to very low and cover, leaving a gap for steam to escape and simmer very gently for about 3 hours, skimming off any scum that rises to the surface. After 2½–2¾ hours of this cooking time, add the stock cubes; if you prefer a very plain taste, just add a little salt and some freshly ground pepper. The beans should be meltingly tender but should not have started to disintegrate.

Remove and discard the celery stalk; ladle one third of the beans into a bowl, put the rest through the vegetable mill with a fine-gauge disc and return to the saucepan; add the reserved, whole beans and any liquid they have with them. This recipe can be prepared 24 hours in advance up to this point.

Shortly before serving, bring a large saucepan of salted water to the boil and cook the pasta in it until only just tender. Reheat the beans slowly over a low heat while the pasta is cooking; add the drained pasta and then stir in 2 tbsp extra-virgin olive oil, 3 tbsp vinegar, a pinch of salt and plenty of freshly ground black pepper. Mix 1½–2 tsp of the pili-pili relish with 1 tbsp olive oil and a pinch of salt in a small bowl and stir into the beans and pasta mixture.

Ladle into individual soup bowls, drizzle a little more olive oil with the surface, finish with a generous sprinkling of freshly ground black pepper and serve, handing round the salad rocket and a small bowl of pili-pili sauce separately.

Serve with thick slices of granary bread.

Beans Creole with Rice and Fried Bananas

1 lb 6 oz (650 g) dried black beans
1 medium-sized onion, peeled and finely chopped
1 clove garlic, peeled and finely chopped
7 oz (200 g) finely chopped green peppers
1 green chilli pepper, finely chopped
3 tbsp olive oil
salt and pepper
For the rice:
10 oz (300 g) par-boiled Patna rice
scant 1½ pints (800 ml) chicken stock (see page 37)
½ onion
½ green pepper
3½ fl oz (100 ml) dry white wine
1½ oz (45 g) butter
4 tbsp chopped coriander
1 lime, quartered
For the fried bananas:
5 bananas
plain flour
sunflower oil
ground cinnamon

Place the beans in a sieve and rinse well under the cold tap, drain and place them in a large, heavy-bottomed flameproof casserole dish with 3½ pints (2 litres) water. Bring slowly to the boil, cover, leaving a gap for steam to escape, reduce the heat to very low and simmer gently for 2 hours, stirring occasionally. When the beans have been cooking for 1¾ hours add a large pinch of salt.

Heat 3 tbsp oil in a large frying pan and gently fry the chopped onion, garlic, peppers and fresh chilli pepper over a low heat for 10 minutes, stirring frequently. Add a little salt and pepper, stir and then mix into the beans. Simmer the beans for a further 15–20 minutes, so that they are very tender.

During this time, cook the rice and fried bananas. Have the stock at a gentle boil. Peel, trim and finely chop the ½ onion and ½ pepper and fry them in 1 oz (30 g) butter over a moderately low heat for 5 minutes, stirring frequently. When they are pale golden brown, add the rice and cook for 1 minute while stirring; pour in the wine and continue cooking until it has completely evaporated. Gradually pour in the boiling stock, stirring as you do so; cover and simmer gently over a low heat for 15 minutes, or until the rice is tender but still firm and has absorbed all the liquid. Remove from the heat and leave to stand, covered, for 5 minutes, then use a fork to gently stir the rice, separating the grains. Stir in the coriander and ½ oz (15 g) solid butter.

While the rice is simmering you can cook the bananas. Peel and cut them lengthwise in half; cut each half into 1½ in (4 cm) pieces. Coat these all over with flour, shaking off excess. Heat plenty of oil in a frying pan and when it is very hot but not smoking, add the bananas and fry briefly, until they begin to crisp and brown on the outside; remove from the oil and drain briefly on kitchen paper in a warm oven.

Transfer the rice to a heated serving dish; arrange the bananas all round the edge and sprinkle them with a little cinnamon. Serve the beans straight from their cooking dish and garnish each helping with a fresh lime wedge for squeezing over the rice.

—— • ——

COUNTRY LENTIL SOUP

1 lb (500 g) continental lentils

1 green celery stalk

1 large clove garlic

3 or 4 sage leaves

4 tinned anchovy fillets

1 14-oz (400-g) tin tomatoes

4 thick slices bread, baked until crisp

3½ oz (100 g) grated hard cheese

2 tbsp extra-virgin olive oil

½ oz butter

salt

Place the lentils in a sieve and rinse very thoroughly under running cold water; leave to soak in a bowl of cold water for 1 hour. Heat 3½ pints (2 litres) water in a large, heavy-bottomed flameproof casserole dish and as soon as it reaches boiling point, add the lentils. Boil them gently for 1 hour.

While they are cooking, finely chop the celery, garlic and sage and fry gently in the oil and butter for 10 minutes over a very low heat, stirring frequently. Remove from the heat, add the anchovy fillets and crush them into the oil and butter, using the back of a wooden spoon. Return the frying pan to a low heat when they have broken up. Add the tomatoes and cook uncovered for 5 minutes, stirring to reduce and thicken. When the lentils are tender, use a ladle to remove approx. 12 fl oz (350 ml) of the cooking liquid from the casserole dish. Stir the tomato mixture into the lentils and allow to return to a very gentle boil. Add a little salt to taste. Place the crisp, dry bread slices in the bottom of the soup bowls and sprinkle the cheese over them; ladle in the soup and serve immediately.

—— • ——

TUSCAN CHICK PEA SOUP

¾ lb (350 g) chick peas

3½ pints (2 litres) vegetable stock (see page 38) or chicken stock (see page 37)

1 sprig rosemary, bound with kitchen string (see method)

2 cloves garlic, unpeeled

1 tbsp concentrated tomato purée

1 tbsp bicarbonate of soda

For the relish:

3 cloves garlic, peeled and crushed

3–4 tbsp extra-virgin olive oil
1 red chilli pepper, finely chopped
3 tinned anchovy fillets
½ medium-sized tin tomatoes
salt and black peppercorns
Serve with:
mixed salad
wholemeal bread

Rinse the chick peas and soak overnight in a large bowl of cold water with the bicarbonate of soda. The next day, drain them and place in a large, heavy-bottomed flameproof casserole dish with the stock, the rosemary (unless you wind kitchen string round it tightly or tie it up in a small muslin bag the tough leaves will come off during cooking and spread through the soup; use a pinch of powdered rosemary if preferred). Add the unpeeled garlic cloves and the tomato purée.

Bring slowly to the boil, skimming off all the scum that rises to the surface. Position the lid so that there is a small gap for the steam to escape and simmer very gently for 2¼ hours, or until the chick peas are very tender. Take about two 8 fl oz (250-ml) measuring cups of the chick peas out of the casserole dish and put through a food mill with a fine-gauge disc fitted. Remove the rosemary bundle and the garlic cloves. Return the puréed chick peas to the casserole dish. Add salt to taste.

Shortly before serving the soup, reheat it gently over a low heat. Make the relish: gently fry the 3 crushed garlic cloves and the chilli pepper in 4 tbsp oil in a non-stick frying pan for 1 minute while stirring; remove from the heat and add the anchovy fillets. Crush these into the oil with the back of a wooden spoon. When they have broken up, stir in the tomatoes and cook over a moderate heat for 3–4 minutes, stirring to reduce.

Mix the contents of the frying pan into the chick pea soup. Serve with a generous sprinkling of freshly ground pepper.

Served with a mixed salad and wholemeal bread, this soup makes a very sustaining meal.

PEAS AND FRESH CURD CHEESE, INDIAN STYLE
Panir mater

1 lb (450 g) frozen peas, thawed
1 large onion, peeled and finely chopped
2 cloves garlic, peeled and finely chopped
1¾-in (2-cm) piece fresh ginger, peeled and grated
1 green chilli pepper, finely chopped
2 large ripe tomatoes, skinned, seeded and chopped
1 tsp cumin seeds
1 tbsp chopped coriander
4½ fl oz (140 g) sunflower oil
pinch saffron
salt
For the fresh curd cheese or panir *(makes approx. 6 oz (180 g):*
3½ pints (2 litres) full-cream milk
1 tsp salt
1–2 tbsp wine vinegar or cider vinegar

Make the curd cheese. Bring the milk slowly to the boil in a large saucepan. Stir in the salt, reduce the heat to very low and gradually add the vinegar to the gently simmering milk while stirring continuously, pouring in a very little at a time. Stop adding vinegar when the milk curdles and thickens. Line a sieve with a piece of muslin and pour the curdled milk into it. Gather up the edges of the muslin cloth, twist them round and force out all the liquid from the curd. Place this fresh curd cheese, still in the muslin, in a large bowl and flatten slightly by hand, place a small plate on the curd, with a weight on top and leave to stand for 2 hours. Drain off the liquid (whey), unwrap the cheese and place on a pastry board. Cut into dice.

While the cheese is being pressed, prepare all the vegetables. Heat 6 tbsp of the oil in a large non-stick frying pan and fry the diced curd cheese until pale golden brown. Take them out of the pan and drain on kitchen paper.

Heat 3 tbsp oil in a large, heavy-bottomed saucepan or flameproof casserole dish and fry the onion gently for 15 minutes, stirring frequently. When the oil is pale golden brown, add the ginger and chilli pepper and cook gently while stirring for about 30 seconds. Add the chopped tomato, the peas and a pinch of salt, cover and cook for 20 minutes over a very low heat, until the peas are tender, adding a little hot water when necessary to moisten. Add the cheese, cook gently for just long enough to heat them through and then draw aside from the heat. Add a little more salt if needed. Transfer to a serving dish, sprinkle with chopped coriander and serve very hot, with hot chapatis.

———•———

MILANESE EGGS AND PEAS

10 oz (300 g) peas, fresh or frozen
2 large spring onions, outer layer removed
1 oz (30 g) butter
1 slice cooked ham, approx. ¼ in (½ cm) thick
5 fl oz (150 ml) vegetable stock (see page 38)
4 eggs
salt and pepper
Serve with:
Potato Mousseline (see page 197)

Slice the trimmed spring onions into thin rings and sweat in the butter for 10 minutes.

Chop the ham coarsely, add to the frying pan and cook gently, stirring and turning, for 1 minute. Add the peas, stock and a little salt and pepper. Bring to a gentle boil, cover and simmer gently for 25 minutes, or until the peas are tender. Add a little more salt and pepper and continue cooking over a slightly higher heat until the liquid has evaporated.

Reduce the heat to very low and with the back of a wooden spoon, make 4 evenly spaced, fairly large, shallow depressions in the contents of the frying

pan and break an egg into each one. Season the eggs with salt and pepper. Cover and cook for 2–3 minutes or until the whites have set. If you like your egg yolks partially set, continue cooking for another 1–2 minutes. Serve at once.

———•———

PEA MOULDS WITH AUBERGINE MOUSSELINE

3½ oz (100 g) frozen peas
1 small shallot, finely chopped
butter
2 basil leaves
5 fl oz (150 ml) chicken stock (see page 37)
2 egg yolks
2 whole eggs
5 fl oz (150 ml) double cream
5 fl oz (150 ml) milk
1½ oz (40 g) grated Parmesan cheese
nutmeg
salt and pepper
For the aubergine mousseline:
2 medium-sized aubergines (approx. 1¼ lb (600 g) total weight)
2 large ripe tomatoes, skinned and seeded
6 basil leaves
2 tbsp lemon juice
6 tbsp olive oil
½ clove garlic, peeled
½ dried chilli pepper, seeds and stalk removed
salt and pepper
Garnish with:
sprigs of basil
Serve with:
French Beans with Crispy Bacon (see page 236)

223

Preheat the oven to 400°F (200°C) mark 6; bake the whole aubergines just as they are for 25–30 minutes, turning them half way through this time.

Make the green pea moulds: sweat the chopped shallot in ½ oz (15 g) butter for 10 minutes in a heavy-bottomed saucepan, then add the peas, basil and the stock. Cover and simmer gently for about 25 minutes or until the peas are very tender, stirring and adding a little more stock now and then. Add a little salt and pepper and continue cooking over a slightly higher heat while stirring to allow all the liquid to evaporate. Put the contents of the saucepan through a vegetable mill with the fine-gauge disc fitted, collecting the purée in a large bowl.

When the aubergines have finished cooking, take them out of the oven and leave to cool completely. Lower the oven to 350°F (180°C) mark 6.

Grease four 7-fl oz (200-ml) timbale moulds or ramekin dishes with butter (or use six 4½ fl oz (140 ml) ones) and bring plenty of water to the boil in the kettle. Beat the 2 egg yolks briefly with the whole eggs, a pinch each of salt, pepper and nutmeg. Lightly beat in the cream, milk, puréed peas and cheese. Add a little more salt and pepper to taste and pour into the moulds; the mixture should come close to their rims. Place them carefully in a roasting tin and pour sufficient boiling water into the pan to come two-thirds of the way up the sides of the moulds. Cook in the oven for 30 minutes. While they are cooking, make the sauce. Peel the aubergines, cut lengthwise into quarters, scoop out the central seed-bearing section if there are any seeds discernible and cut the remaining flesh into small pieces. Place these in the food processor or blender with the coarsely chopped tomatoes, 6 basil leaves, the lemon juice, a generous pinch of salt and plenty of freshly ground pepper, the olive oil, garlic and crumbled chilli pepper. Process at high speed until very smooth and creamy. Pour this thick sauce into a bowl. When the moulds have had 30 minutes in the oven, take them out and leave to stand for 5 minutes before you run the point of a sharp knife round the inside of the moulds to loosen. Turn out on to hot plates, spoon some of the aubergine sauce to one side and garnish each serving with a sprig of basil.

CASEROLE OF PEAS AND BEANS WITH ARTICHOKES

5 oz (150 g) fresh or frozen peas
7 oz (200 g) broad beans, shelled and seed cases removed
6 very young, tender artichoke hearts or 12 fresh artichoke bottoms
juice of 1 lemon
1 small Cos lettuce
1 medium-sized onion
1 clove garlic
1 sprig thyme
4 tbsp olive oil
salt and pepper
Serve with:
wholemeal bread

If you are using frozen peas, partially defrost them; it is best to remove the casing from each broad bean unless they are very young and tiny as they tend to be leathery and tough once cooked. If you are using tender artichokes, see page 18 for preparation; cut the hearts lengthwise in quarters. Older, tougher artichokes should be stripped right down to their dish-shaped bottoms; these can then be cut into quarters or sliced horizontally into 2 or more pieces. In both cases drop each artichoke as soon as it is prepared into a bowl of cold water acidulated with the lemon juice. Separate the lettuce leaves, removing the hard base, wash and dry them and cut into a chiffonade (see page 24). Cut the peeled onion lengthwise into quarters and then slice very thinly.

Drain the artichokes and place in a wide, fairly shallow flameproof casserole dish or saucepan with the other vegetables, the whole but slightly crushed garlic clove and the sprig of thyme. Season with a little salt and pepper. Add 5 fl oz (150 ml) water and the olive oil.

Bring to the boil, cover tightly, turn down the heat and simmer very gently for about 15 minutes or until the vegetables are tender but not at all mushy.

Braised Peas, Artichokes and Lettuce

10 oz (300 g) young peas or petits pois, *fresh or frozen*
2 lettuces
juice of 1 lemon
6 young, tender artichoke hearts (see previous recipe)
1 small onion, peeled and finely chopped
1 sprig thyme
7 fl oz (200 ml) vegetable stock (see page 38)
2 tbsp chopped parsley
scant 1 oz (20 g) butter
2 tbsp olive oil
salt and pepper
Serve with:
mozzarella cheese, dressed with olive oil, chopped fresh basil, salt and pepper; seafood dishes; roast meat

Preheat the oven to 325°F (170°C) mark 3. Choose a well-flavoured lettuce variety for this recipe; the iceberg type is not suitable. Trim the lettuces, remove and discard the outer leaves, remove the base, separate and wash the inner leaves, drain well and cut into a chiffonade (see page 24). Have a bowl full of cold water acidulated with the lemon juice standing ready; as you prepare the artichokes, drop them into it to prevent discoloration. If you are using artichoke bottoms, allow 12.

Sweat the onion in the butter and oil in a large, shallow flameproof casserole dish with a tight-fitting lid with the sprig of thyme for about 10 minutes, stirring frequently. Add the well drained artichokes, turn up the heat a little and fry them for 15 minutes, stirring occasionally. Add the peas, season lightly with salt and freshly ground pepper and cook with the lid on over a low heat for 5 minutes. If using frozen peas, cook for several minutes longer, until they have thawed completely. Add the shredded lettuce and the stock, cover tightly again and place in the oven to cook for 1 hour. When this time is up, correct the seasoning, sprinkle with the parsley and

serve directly from the casserole dish or transfer to a heated serving plate.

———— • ————

Cantonese Mangetout and Prawns

1 lb (500 g) mangetout or sugar peas
1 lb (500 g) tiger prawns, net weight, heads removed
5 fl oz (150 ml) shellfish stock (see page 38)
1 ¾-in (2-cm) piece fresh ginger, finely chopped
4 tbsp plain flour or cornflour
4 spring onions, sliced
1 clove garlic, peeled and crushed
2 fl oz (60 ml) Chinese rice wine or dry sherry
6 tbsp sunflower oil
salt
Serve with:
Cantonese Stir-fried Rice and Peas (see page 233)
Bamboo Shoot and Asparagus Soup (see page 118)

Peel the prawns; make a neat incision near the end of the tail to get to the black 'vein' or intestinal tract running up the back of each crustacean; lift this up so that you can pull it out cleanly.

Place the prawns in a bowl and add 1 tsp salt and the finely chopped ginger, using your fingertips to mix thoroughly. Leave for 30 minutes, then mix in the flour or cornflour, using your fingertips again, to coat the prawns.

Heat a large saucepan of salted water in which to cook the mangetout; prepare these as described on page 22, wash, drain and blanch for 2 minutes in the boiling, salted water. Drain in a sieve or colander and refresh, holding the sieve under running cold water to rinse thoroughly. Drain once more. Cut the inner part of the spring onion (both bulb and leaves) into 1½-in (4-cm) lengths and then cut these lengthwise, into julienne strips (see page 24).

Heat 3 tbsp of the oil in a wok or large frying pan and when it is very hot, add the garlic clove (left whole, but crushed with the flat of a large knife blade) and the mangetout and stir-fry for 3–4 minutes. Remove from the wok with a slotted spoon and keep warm on a hot plate while you cook the prawns.

Add the remaining oil to the wok and stir-fry the spring onions for 1 minute; add the prawns, leaving behind any excess floury mixture in the bowl and stir-fry for 2 minutes. Sprinkle with the rice wine and approx. 4 fl oz (120 ml) of the shellfish stock and stir as the sauce thickens; add the mangetout, reduce the heat and continue stirring for 1 minute over a lower heat. Simmer for a few seconds longer, then remove from the heat and serve immediately, with Cantonese Stir-fried Rice and Peas. The Bamboo Shoot and Asparagus Soup can be served before or after this dish as part of a Chinese meal.

BEANS AND EGGS MEXICAN STYLE
Frijoles con huevos rancheros a la mexicana

14 oz (400 g) dried red kidney or pinto beans
1 small onion, peeled and finely chopped
10 oz (300 g) tinned tomatoes, drained and chopped
1–2 fresh or dried red chilli peppers, finely chopped
1 small garlic clove, peeled and finely chopped
scant 1 oz (20 g) butter
2 tbsp olive oil
salt and pepper
For the eggs:
1 medium-sized onion, peeled and finely chopped
½–1 fresh green chilli pepper, finely chopped
10 oz (300 g) tinned tomatoes, drained and chopped
4 eggs
2 tbsp olive oil
salt and pepper

Serve with:
1 ripe avocado, peeled and sliced
tortilla chips

Rinse the beans thoroughly in a sieve under running cold water and place in a large, heavy-bottomed flameproof casserole dish with 2¾ pints (1½ litres) water, half the chopped onion, 4 tbsp of the chopped tomatoes and the chilli pepper; bring to the boil. Skim off any scum that rises to the surface; reduce the heat to low, cover, leaving a gap for the steam to escape and simmer for 2½ hours or until the beans are very tender, stirring now and then. Season with salt and pepper. Sweat the remaining chopped onion with the garlic in the butter and olive oil for 10 minutes over a very low heat, then add the remaining chopped tomato and a pinch of salt, increase the heat to high and cook, uncovered, for 3 minutes. Add about 8 oz (225 g) of the cooked beans and crush them coarsely with a fork or the back of a wooden spoon. Stir well, add another, similar quantity of cooked beans and crush these. Continue doing this until all the beans have been added. Stir while cooking for a few minutes, to thicken the mixture. Cover and keep warm.

Cook the eggs: sweat the onion until tender but not at all coloured with the chilli pepper in a wide non-stick frying pan with 2 tbsp oil. Add 10 oz (300 g) chopped tomatoes and a little salt and pepper to taste. Boil, uncovered, over a fairly high heat for 2–3 minutes, stirring to reduce and thicken. Make 4 hollows or spaces with the back of a wooden spoon and break the eggs into them; season the eggs with salt and pepper, turn down the heat, cover and cook gently for 2–3 minutes.

WARM BEAN SALAD

1 lb (500 g) fresh or dried white haricot beans
1 small onion, peeled
1 medium-sized carrot, peeled

2 bay leaves
1 sprig rosemary, bound with kitchen string (see method) or pinch powdered rosemary
2 thick rashers streaky bacon, chopped
1 tbsp chopped parsley
olive oil
wine vinegar
salt and black peppercorns

If using dried beans, soak them in plenty of cold water with 1 tsp bicarbonate of soda overnight. Drain, then rinse well. If using fresh beans, simply rinse and drain. Place them in a large, heavy-bottomed flameproof casserole dish with the onion, carrot, the bay leaves and the sprig of rosemary, with kitchen thread wound tightly round it to prevent the tough leaves dropping off during cooking. Add a few whole peppercorns and the bacon. Tie the rosemary and peppercorns in a small piece of muslin if preferred. Add sufficient water to cover all the ingredients easily and bring to the boil, skimming off the scum that rises to the surface. Cover with the lid, leaving a small gap to allow the steam to escape and simmer gently over a low heat for about 2 hours, stirring now and then, until the beans are very tender. Add salt to taste just before they are done.

Use a slotted spoon or ladle to transfer portions of the beans to individual bowls or deep plates; spoon a little of the liquid over them. Sprinkle sparingly with wine vinegar, with plenty of freshly ground black pepper, the chopped parsley, and a little oil.

———•———

FRENCH BEAN AND POTATO PIE

1 lb (450 g) fresh or frozen French beans
1 lb (500 g) floury potatoes
1 large onion, peeled and finely chopped
2 eggs
2 oz (50 g) grated Parmesan cheese

scant 1 tbsp chopped fresh marjoram leaves
1 clove garlic, peeled and grated
2 oz (60 g) sliced Mortadella
2 oz (50 g) butter
2 tbsp olive oil
approx. 3 oz (90 g) fine dry breadcrumbs
nutmeg
salt and pepper
Serve with:
green salad
tomato salad

Wash the potatoes under running cold water but do not peel them; steam for 30–40 minutes until very tender. Boil the trimmed beans in salted water for 20–25 minutes, until tender. Drain. Preheat the oven to 400°F (200°C) mark 6. Sweat the onion in 1 oz (30 g) of the butter and 1 tbsp of the oil for 20–25 minutes over an extremely low heat, stirring frequently, until it has almost disintegrated and is pale golden brown. As soon as the potatoes are done, spear them with a carving fork and peel them while they are still boiling hot; put through the potato ricer twice or push through a fine sieve into a large bowl.

Make sure you have drained every drop of water from the beans, place them in the food processor and process to a smooth purée. Mix this purée with the potatoes, adding the eggs and onion. Place the Parmesan, marjoram, garlic and Mortadella in the food processor and process until completely blended; stir into the bean and potato mixture, seasoning with a little salt, plenty of freshly ground pepper and a pinch of nutmeg.

Oil a 9½-in (24-cm) shallow cake tin or quiche dish (at least 1½ in (4 cm) deep) and fill it with the prepared mixture, levelling the surface with a palette knife. Sprinkle the top with a layer of breadcrumbs, dot small flakes of the remaining butter over the surface and bake in the oven for 10 minutes; place under a very hot grill for about 3 minutes to crisp and brown the surface.

CANNELLINI BEANS WITH HAM AND TOMATO SAUCE

14 oz (400 g) dried cannellini or haricot beans

1 sprig rosemary, bound with kitchen string or pinch powdered rosemary

1 garlic clove, unpeeled

2 spring onions, finely chopped

1 small clove garlic, peeled and finely chopped

1 3½–4 oz (100-g) slice fairly fatty cooked ham

2 tbsp chopped fresh basil

2 tbsp chopped parsley

1 lb (500 g) ripe tomatoes, peeled, seeded and chopped

2 tbsp olive oil

salt and pepper

Rinse the beans in a sieve under running cold water, place in a large flameproof casserole dish, add sufficient cold water to cover them by about 2 in (5 cm), bring slowly to the boil then simmer over a moderate heat for 3 minutes. Drain in a sieve and rinse well under running cold water; drain again and replace in the casserole dish, covering amply with water once more. Add the sprig of rosemary, wound round tightly with black thread or thin kitchen string to prevent the tough leaves falling off during cooking; add the first garlic clove, whole and still in its skin. Bring to the boil, turn down the heat to very low and cover, leaving a gap for the steam to escape. Simmer gently for 2 hours or until the beans are tender; add a little salt after about 1¾ hours.

Make the ham and tomato sauce: heat 2 tbsp olive oil in a very large heavy-bottomed saucepan or flameproof casserole dish and gently fry the spring onions, the second, peeled and finely chopped clove of garlic and the finely chopped lean and fat of the ham for 10 minutes, stirring frequently. Add the chopped tomatoes and a little salt and pepper. Turn up the heat and cook, uncovered, for about 15 minutes, stirring frequently to allow the sauce to reduce and thicken.

Drain the cooked beans, reserving about 4 fl oz

(120 ml) of their cooking water, and add to the tomato sauce, together with some of the reserved cooking water. Stir over a very low heat for a few minutes, adding more cooking water if necessary.

———— • ————

NORTH AFRICAN VEGETABLE CASSEROLE

10 oz (300 g) dried chick peas

2 large aubergines

3 onions

2¼–2½ lb (1 kg) ripe tomatoes

bicarbonate of soda

light olive oil

salt and pepper

Soak the chick peas overnight in a large bowl of cold water. The next day, drain them and place in a large, flameproof casserole dish; add sufficient cold water to cover them easily and a pinch of bicarbonate of soda. Bring to the boil quickly over a high heat, then reduce the heat to very low; cover, leaving a small gap for the steam to escape. Simmer gently for 2 hours or until they are tender, topping up the water level with some more boiling water if necessary. Add salt when the chick peas have nearly finished cooking. When they are done, drain and set aside.

Preheat the oven to 350°F (180°C) mark 4. Peel off lengthwise strips of skin from the aubergines approx. ½ in (1 cm) wide, leaving strips of the same width between them; cut the aubergines lengthwise into quarters and then cut these sections across, into pieces about 1½ in (4 cm) long. Heat sufficient light olive oil to amply cover the bottom of a medium-sized non-stick frying pan and when hot fry the aubergine pieces in batches over a moderately high heat, stirring them until they are lightly browned. Remove from the frying pan with a slotted spoon or spatula and finish draining on kitchen paper; sprinkle with a little salt and pepper. Aubergines absorb a great deal of oil, you may need to add more to fry the later batches.

Slice the peeled onion thinly. Fry gently for 10 minutes, stirring in the oil left over from frying the aubergines; when the onion is soft and very pale golden brown, remove from the heat.

Blanch the tomatoes in boiling water for 10 seconds, peel them, cut into quarters and remove all the seeds and any tough parts. Chop coarsely, transfer to a bowl and season with salt and pepper. Spread out the aubergine pieces in a single layer in a shallow, flameproof casserole dish. Cover with the onions, sprinkle with a little salt and pepper, cover the onion layer with the chick peas and top with a layer of tomatoes. Pour in 10 fl oz (300 ml) water to one side of the dish, bring to a gentle boil and bake for about 45 minutes. Serves 6.

———— • ————

GREEK CHICK PEA CROQUETTES
Pittarúdia

7 oz (200 g) dried chick peas
15 oz (425 g) tinned tomatoes, drained and chopped
1 small onion, peeled and grated
1½ tbsp ground cumin
1½ tbsp dried mint
scant 1 tsp freshly ground black pepper
4 oz (120 g) plain flour
1 tsp baking powder
4 tbsp fine breadcrumbs
1½ tbsp chopped parsley
light olive oil
salt
Serve with:

Tsatsiki (see page 149)
hot pitta bread

Soak the chick peas in plenty of cold water overnight or for at least 12 hours. Drain thoroughly, transfer to a food processor and process until coarsely chopped. Transfer to a large mixing bowl and stir in the chopped tomato, the grated onion, cumin, mint, pepper and 1 tsp salt. Sift in the flour and baking powder together and stir in the fine soft breadcrumbs and parsley. The mixture should be soft, but capable of holding its shape when drawn into a peak. Leave to rest for 30 minutes.

Make the Tsatsiki (see page 149) and chill.

Pour enough olive oil into a very wide non-stick frying pan to form a layer about ¾ in (2 cm) deep or heat plenty of oil in a deep fryer to 350°F (180°C). When the oil is very hot but not smoking, drop heaped tablespoonfuls of the chickpea mixture into it, making sure they do not touch one another and frying a few at a time; if shallow-frying turn once when the underside is crisp and golden brown (after about 1 minute); the croquettes should puff up as they cook. Remove the first batch from the oil with a slotted spoon and finish draining on kitchen paper on a hot plate while you fry the other batches. Serve as soon as they are all cooked, with the Tsatsiki and hot pitta bread.

———— • ————

MUNG DHAL FRITTERS

6 oz (180 g) mung dhal (dried split mung beans)
5 small green peppers, finely chopped
½–1 green chilli pepper, finely chopped
generous pinch baking powder
3 tbsp Greek yoghurt
1 thin piece fresh ginger, peeled and finely chopped
3½ oz (100 g) cooked green peas
1½ tbsp ghee (clarified butter, see page 36)

1 tsp cumin seeds

2 tbsp chopped coriander leaves

4 tbsp sunflower oil

salt and pepper

Soak the mung dhal (yellow or cream Indian pulses) in a large bowl of cold water overnight. Drain the next day. Remove the stalk, seeds and inner membrane from 3 of the peppers, and process with the mung dhal in a food processor or blender with 2 tbsp cold water until they form a thick, soft purée. Stir in the baking powder, yoghurt and a pinch of salt. Make the green pea filling. Mash the peas coarsely with a fork or a potato masher. Chop the ginger with the remaining 2 peppers and the remaining chilli pepper. Heat the ghee in a small frying pan and stir-fry the cumin seeds over a gentle heat for a few seconds. Add the peppers, chilli and ginger, fry gently for 1 minute, stirring, then add the mashed peas and a small pinch of salt. Stir over a low heat for 1 minute. Add the chopped coriander, mix well and remove from the heat. Heat 3 tbsp oil in a 6-in (15-cm) non-stick omelette pan; when it is very hot, spoon 3–4 tbsp of the lentil mixture into the pan; spread the mixture out with a non-stick spatula to form a thick round cake about 2½–3 in (7–8 cm) in diameter; make a small hole in the centre before the cake has time to set. Place 1 heaped tsp of the filling in the centre of the cake, and press to flatten. Fry for 2 minutes then turn and fry for a further 1–1½ minutes. Take out of the frying pan, keep hot on kitchen paper while you repeat the procedure until all the lentil mixture and pea filling is used up.

———•———

RED LENTILS WITH RICE INDIAN STYLE

7 oz (200 g) red lentils

3 thin pieces fresh ginger, peeled

½ tsp ground turmeric

1 tsp coriander seeds

3 tbsp ghee (clarified butter, see page 36)

1 tsp cumin seeds

2 tbsp chopped coriander leaves

generous pinch cayenne pepper or chilli powder

salt

For the rice:

10 oz (300 g) Basmati rice

1 tbsp butter

salt

For the saffron butter:

generous pinch saffron threads

4 tbsp ghee (clarified butter, see page 36)

Serve with:

chapatis (see page 81)

Cook the rice by the absorption method: measure out just under 14 fl oz (390 ml) cold water. Rinse the rice in a sieve under running cold water until the water runs out clear. Drain, place in a bowl, add sufficient cold water to cover and leave to soak for 20 minutes, then drain again. Spread the rice out in the bottom of a heavy-bottomed saucepan or flameproof casserole dish, add the measured quantity of cold water, ½ tsp salt and the butter. Heat until the water starts to boil, cover tightly and cook gently over a very low heat for up to 20 minutes.

While the rice is soaking then cooking, prepare the lentils. Rinse them well under running cold water; drain and place in a saucepan or flameproof casserole dish with 1¼ pints (700 ml) cold water and bring to the boil, skimming off any scum that rises to the surface. Add the slices of fresh ginger and stir in the turmeric. Reduce the heat to low, cover, leaving a chink through which the steam can escape and simmer gently for 20 minutes until the lentils are tender; stir occasionally to prevent them catching and burning on the bottom of the saucepan or casserole dish. Remove and discard the pieces of ginger, add salt to taste, stir and remove from the heat. Pound the coriander seeds with a pestle and mortar (or grind in the food processor) and sift to get rid of the tough seed cases. Heat 3 tbsp ghee in a small saucepan and fry the cumin seeds for a few seconds. Add the chopped coriander leaves and the cayenne pepper or chilli powder, stir well and then mix with the lentils. Add a little more salt if necessary, spoon into a heated serving dish and sprinkle with the chopped coriander garnish. Make the saffron butter: place the saffron threads in a small bowl, add 1 tbsp boiling water and stir well, crushing the threads with the back of a wooden spoon. Heat 4 tbsp ghee in a small saucepan, stir in the saffron and water, and mix briefly. Transfer the rice to a heated serving dish and pour the saffron butter all over it. Serve at once, with the chapatis.

———— • ————

BORLOTTI BEAN PURÉE

1¼ lb (600 g) dried borlotti beans

1 sprig fresh rosemary, bound tightly with kitchen string

1 clove garlic, left whole and unpeeled

salt

5 fl oz (150 ml) crème fraîche *or double cream*

1 oz (30 g) unsalted butter

salt and pepper

Cook the beans by the quick method described in the recipe for Canellini Beans with Ham and Tomato Sauce on page 228. When cooked, drain, reserving about 8 fl oz (225 ml) of the cooking water. Alternatively, use 2 medium-sized tins borlotti beans and reserve the liquid when you drain them. Put the beans through the vegetable mill with a fine-gauge disc fitted; reheat the purée, adding some or all of the reserved liquid; it should be very thick but not stiff. Stir continuously as you heat it over a very

low heat and when it comes to the boil, keep stirring as you gradually pour in the cream a little at a time. Continue cooking and stirring for 1 minute after all the cream has been added, to thicken, then take off the heat. Add a little salt and plenty of freshly ground pepper to taste and stir in the solid unsalted butter until it has melted completely.

———— • ————

MEXICAN BEANS WITH GARLIC AND CORIANDER

1¼ lb (600 g) soaked red kidney or pinto beans

1 piece fresh ginger, unpeeled

salt

For the sauce:

4 garlic cloves, peeled and sliced

5 tbsp olive oil

2 medium-sized ripe tomatoes or tinned tomatoes, drained and chopped

1 tsp pili-pili hot relish (see page 113) or chilli powder

1½ tbsp chopped coriander leaves

salt and pepper

Serve with:

green salad

hot pitta bread

Place the beans in a flameproof casserole dish with sufficient water to completely cover them and the unpeeled piece of ginger. Bring to the boil and then simmer for 2½ hours, partially covered to allow the steam to escape. The beans should be very tender. Add salt when they are nearly done. Drain, reserving about 8 fl oz (225 ml) of the cooking water. Discard the ginger. Put them through the vegetable mill with a fine-gauge disc fitted and return the purée to the casserole dish; stir in some or all of the reserved liquid so that the purée is thick but not stiff. Set aside, ready for reheating. Make the sauce: slice the peeled garlic cloves and blanch for 3 seconds in a small saucepan containing some boiling salted water; drain at once and rinse in a sieve under running cold water to refresh. Blot dry with a piece of kitchen paper.

Heat the oil in a heavy-bottomed saucepan and fry the garlic slices very gently over a low heat until they are a very pale golden brown. Increase the heat slightly and add the diced flesh from the skinned, seeded tomatoes; cook for 1 minute while stirring gently. Add a pinch of salt and remove from the heat; stir in the pili-pili relish and keep warm.

Heat the bean purée over a low heat, stirring continuously until it comes to the boil. Season. Spoon the tomato mixture into the middle of each bowl of bean purée, distributing an even number of garlic slices on each serving. Sprinkle with the chopped coriander and serve with hot pitta bread and rocket salad.

———— • ————

BROAD BEANS WITH CREAM

1¼ lb (600 g) shelled broad beans
1 clove garlic, peeled and partially crushed
9 fl oz (250 ml) crème fraîche *or double cream*
1 oz (30 g) butter
salt and pepper

Heat plenty of salted water in a large saucepan. When it comes to the boil, add the broad beans, turn down the heat and cover, leaving a small gap through which the steam can escape. Simmer for 10 minutes. Drain the beans in a colander, refresh by rinsing them under running cold water, then remove the seed cases, breaking up the bright green cotyledons inside as little as possible.

Heat the butter in a wide, heavy-bottomed frying pan with the garlic clove in it, add the beans and cook over a gentle heat for 2 minutes, stirring gently. Add the *crème fraîche* or double cream and cook uncovered over a moderate heat for 3–4 minutes. Season and serve at once.

JAPANESE STIR-FRIED RICE WITH GREEN PEAS AND MIXED VEGETABLES
Yakimeshi

2 oz (60 g) cooked green peas
2 spring onions
½–1 green chilli pepper, sliced into rings
1 young tender carrot, diced
½ red pepper, diced
½ cucumber
1 egg
generous pinch dashinomoto (instant fish stock)
3½–4 oz (100 g) diced cooked chicken or pork
10 oz (280 g) rice
pinch caster sugar
4 tbsp sunflower oil
salt
Serve with:
Miso and Tofu Soup (see page 217)

Cook the rice in advance; allow to cool quickly, then refrigerate in a covered container until 30 minutes before you plan to serve this dish.

Prepare all the vegetables: cut off the skin of the cucumber together with a ¼-in (½-cm) layer of the flesh immediately beneath it and dice these thick strips. Reserve the remaining inner part of the cucumber for other dishes. Beat the egg briefly in a small bowl with the *dashinomoto* (if you do not have any of this instant Japanese fish stock, use a pinch of crumbled chicken stock cube, 2–3 tbsp cold water and a pinch each of sugar and salt).

Heat 1 tbsp oil over a low heat in a wide non-stick frying pan and pour in the beaten egg, tipping it this way and that so that it spreads out into a very thin omelette; cook gently until it has completely set but do not allow to brown. Remove the omelette from the frying pan, spread out on a chopping board and leave to cool before chopping it finely.

Heat the remaining oil in the wok and stir-fry all the vegetables over a high heat for 2 minutes. Add a pinch of salt, followed by the rice. Reduce the heat and break up any lumps of rice formed by the grains sticking together while mixing the rice with the vegetables for about 3 minutes, by which time it should be hot. The vegetables should still be very crisp. Sprinkle with another teaspoon of salt and mix in the chopped omelette. Take the wok off the heat and serve. Serve Miso and Tofu Soup (see page 217) at the same meal.

———— • ————

PURÉE OF PEAS

12 oz (350 g) frozen peas
2 medium-sized floury potatoes
1 shallot, peeled and finely chopped
3 fresh basil leaves
1 pint (600 ml) vegetable stock (see page 38)
1 egg yolk
½ clove garlic, peeled and grated
1½ oz (40 g) freshly grated hard cheese
2 tbsp olive oil
nutmeg
salt and pepper
Serve with:
hot vegetable moulds or egg dishes

Peel and dice the potatoes; place in a large, heavy-bottomed saucepan with the peas, shallot, basil leaves and stock. Bring to the boil, cover and simmer over a low heat for about 30 minutes or until the potatoes are very tender. Drain off and reserve the liquid remaining in the saucepan; purée the cooked mixture by putting it through the vegetable mill with a fine-gauge disc fitted, or rub through a sieve. Return the purée to the saucepan; stir in 2–3 tbsp of

the reserved liquid and heat to just below boiling point over a lot heat. Beat the egg yolk into the purée for a few seconds only, just enough to blend well, using a balloon whisk. Remove from the heat, season to taste with a little salt, some freshly ground pepper and a pinch of nutmeg; stir in the garlic and then the grated cheese. Keep beating as you add the olive oil a very little at a time. Serve at once.

———— • ————

CANTONESE STIR-FRIED RICE AND PEAS

3 oz (90 g) cooked green peas
2 green celery stalks, trimmed and diced
4 large spring onions, thinly sliced
10 oz (300 g) rice
3 eggs
1 thick slice cooked ham weighing approx. 4 oz (120 g), diced
5 tbsp sunflower oil
pinch caster sugar
pinch monosodium glutamate (optional)
salt

Steam or boil the rice 24 hours before you plan to serve this dish, place in a sealed container when cold and refrigerate until 30 minutes before using it. Beat the eggs with a pinch of salt, sugar and monosodium glutamate (if used). Heat 1 tbsp of the oil in a large, non-stick frying pan over a low heat and make a very large, thin omelette; it should be completely set but not at all browned. Remove from the pan, spread out on a chopping board and when completely cold, chop it coarsely. Set aside.

Heat the remaining oil in a large wok and stir-fry the sliced spring onions and the diced celery for 1 minute. Add the peas and the diced ham and stir-fry for 1½ minutes. Add a small pinch of salt and the cold rice; reduce the heat and stir-fry for a few minutes, separating any rice grains that stick together. Sprinkle with a generous pinch of salt and continue to stir-fry for a further 3 minutes, or until the rice has heated through. Add the chopped omelette, mix evenly with the rice and vegetables and remove from the heat.

This dish makes a good foil for other, more highly flavoured Chinese dishes. Serve with soy sauce.

———— • ————

PEA AND WILD RICE TIMBALES

3 oz (80 g) peas, fresh or frozen
1 shallot, peeled and finely chopped
1 sprig thyme
1½ oz (35 g) butter
7 fl oz (200 ml) vegetable stock (see page 38)
3½–4 oz (80 g) wild rice
4½ oz (130 g) brown rice
salt and pepper
Serve with:
Catalan Lobster with Peppers (see page 140), Artichoke Fricassée (see page 129) or Pisto Manchego (see page 139)

Place the wild rice in a small, heavy-bottomed saucepan and add 1 pint (550 ml) water. Bring to the boil, cover tightly, reduce the heat to very low and cook for 30 minutes. Remove from the heat, stir in the brown rice and ½ tsp salt, bring to the boil again quickly, cover tightly, reduce the heat to very low again and continue cooking for about 14 minutes more, or until both the wild rice and brown rice are tender but still firm and all the moisture has been absorbed.

While the rice is cooking, sweat the shallot in ½ oz (15 g) of the butter with the sprig of thyme for 5 minutes, stirring. Add the peas and the stock, bring to the boil, cover and then simmer gently for about 25 minutes, by which time the peas should be very ten-

der. Remove and discard the sprig of thyme and stir the peas over a moderate heat until the liquid has considerably reduced and the peas are just moist. Add salt and pepper to taste. Add the peas to the rice, with scant ½ oz (10 g) butter and stir until the butter has completely melted.

Grease 4 small timbale moulds or ramekin dishes with the remaining butter, fill them with the rice and pea mixture, pressing down gently with the back of a wooden spoon to pack in evenly and firmly. Keep hot in a *bain marie* of very hot water for up to 30 minutes if they are not being served immediately.

Turn out on to heated plates, slightly to one side, leaving room for your chosen accompaniment. These timbales go very well with seafood, vegetarian dishes, veal or poultry. A delicious and unusual light meal can be made by mixing 4 egg yolks with a pinch of salt and pepper and a little fresh lime juice and stirring this into the rice mixture before packing into the moulds; cover each mould with foil or greaseproof paper and bake in a preheated oven at 325°F (170°C) mark 3 in a *bain marie* for about 15 minutes, then serve with bananas lightly fried in butter, sprinkled with a little cinnamon and garnish with lime wedges and sprigs of coriander.

———— • ————

PEAS WITH BACON AND BASIL

10 oz (300 g) frozen peas or petit pois
2 shallots, peeled and finely chopped
scant 1 oz (20 g) butter
4 rashers streaky bacon
7 fl oz (200 ml) vegetable stock (see page 38)
2 basil leaves
1½ tbsp chopped basil
3½ fl oz (100 ml) crème fraîche or double cream
salt and pepper

BEAN AND PEANUT PILAF

3 tbsp dried black-eyed beans
3 tbsp dried red kidney beans or pinto beans
3 tbsp dried chick peas
3 tbsp unsalted whole raw peanuts
14 oz (400 g) long-grain rice (uncooked weight)
6 tbsp ghee (clarified butter, see page 36)
1½ tbsp mustard seeds
1 large onion, peeled and finely chopped
1½ green chilli peppers, seeds removed, finely chopped
2 large ripe tomatoes, skinned, seeded and chopped
1½ tsp mild curry powder or garam masala
3 oz (90 g) bean sprouts
3 tbsp fresh lime juice
salt
Garnish with:
cucumber fans (see page 29)

Serve with:

hot vegetable moulds or timbales; potato pies; roast meat or poultry

Sweat the shallot in the butter in a large, heavy-bottomed flameproof casserole dish for 5 minutes, stirring frequently. Cut off the bacon into thin strips. Add these to the shallot and fry gently over a low heat while stirring for 2–3 minutes. Add the frozen peas, stock, the basil leaves, a small pinch of salt and some freshly ground pepper. When the liquid comes to the boil, cover and simmer gently for 20–25 minutes, stirring occasionally. When the peas are tender, add the cream and cook while stirring to allow it to reduce and thicken a little. Correct the seasoning if needed, stir in the chopped basil and serve immediately.

——— • ———

FRENCH BEANS MAÎTRE D'HOTEL

1¾ lb (800 g) French beans
18 fl oz (500 ml) béchamel sauce (see page 36)
1½ oz (40 g) grated Gruyère cheese
1 tsp chopped fresh mint leaves
2 tbsp chopped parsley
salt
Serve with:
hot vegetable moulds, omelettes, grilled fish or meat

Heat plenty of lightly salted water in a large saucepan. Trim and rinse the beans and add to the boiling salted water to cook, uncovered, until they are tender but still crisp.

While the beans are cooking, make a béchamel sauce of pouring consistency; take off the heat and stir in the cheese. Drain the beans; spread out in a wide, shallow flameproof dish, pour the cheese sauce all over them, heat to a gentle boil and serve, sprinkled with the mint and parsley.

The day before you plan to serve this dish, rinse all the dried beans and the chick peas, place them in a large bowl and add enough cold water to come well above them; soak overnight or for at least 12 hours. Drain, transfer to a large, heavy-bottomed saucepan, add sufficient cold water to cover them and bring to the boil. Skim off any scum that rises to the surface; reduce the heat to low, cover, leaving a small gap through which the steam can escape and simmer for 45 minutes, then add the peanuts and a generous pinch of salt and continue simmering until all the dried beans and chick peas are tender. Drain and set aside. While the beans are cooking, steam the rice and leave to cool at room temperature, then refrigerate in a sealed container until shortly before you start the final cooking stages of the pilaf. The beans can also be refrigerated if you prefer to complete preparation up to this point several hours or a day in advance.

Shortly before serving the pilaf, heat 4 tbsp of the ghee in a very large deep frying pan and fry the mustard seeds gently for a few seconds; when they start to jump about in the pan, add the onion and chilli pepper and fry gently over a low heat for 10 minutes, stirring frequently. Turn up the heat, add the tomatoes (use drained, tinned tomatoes if preferred) and a pinch of salt. Cook, stirring, for 2–3 minutes. Stir in the curry powder or garam masala, the bean sprouts, the cooked beans, chick peas and peanuts and the lime juice. Cover and cook gently over a low heat for 2–3 minutes.

Heat the remaining ghee in another large frying pan and stir-fry the rice for a few minutes, until it is heated through. Sprinkle with a small pinch of salt and combine gently with the vegetables in the larger frying pan. Serve very hot, garnished with cucumber fans.

———— • ————

FRENCH BEANS WITH TOMATOES AND SAGE

1¼ lb (600 g) French beans
1 small onion, finely chopped
2 cloves garlic, peeled and crushed
4 fresh sage leaves
1 lb (500 g) chopped tomato flesh (fresh or tinned)
3 tbsp olive oil
salt and pepper
Serve with:
meat dishes

Trim the beans; cut them into 1¼-in (3-cm) lengths, rinse and drain.

Heat the oil in a large frying pan or fairly shallow saucepan and fry the onion, garlic and sage gently over a low heat for about 10 minutes, stirring frequently. When the onion is just tender, add the beans and chopped tomato, a pinch of salt and some freshly ground pepper, stir well and add just enough cold water to cover the beans.

Bring to a very gentle boil, cover and simmer very gently for about 25 minutes or until the beans are tender but still fairly crisp. Tiny, thin French beans will only take about 15 minutes or less. Check whether there is a lot of liquid left in the pan when the beans are nearly done; if there is an appreciable amount, simmer uncovered for the last 5 minutes' cooking time to reduce. Taste and correct the seasoning; serve very hot. This is a very versatile vegetable dish which can also be served on its own or with crusty bread and goat's cheese as a nourishing light meal.

———— • ————

FRENCH BEANS WITH CRISPY BACON

8 oz (250 g) French beans
16 rashers smoked streaky bacon
scant 1 oz (20 g) butter
salt and pepper
Serve with:
omelettes or vegetable pancakes

Heat plenty of lightly salted water in a large saucepan. Preheat the oven to 325°F (170°C) mark 3. Top and tail the beans, rinse well and boil, uncovered, for 12–15 minutes or until they are tender but still crisp. Drain and leave to cool a little. Grease a wide, shallow ovenproof dish with a little butter. Divide the beans into 8 equal bundles and wrap 2 bacon rashers round each one (do not have the rashers overlapping one another or they will not crisp well). Place these bundles neatly in the dish, dot small flakes of butter here and there over the beans where they are not covered by the bacon and bake in the oven for 10 minutes, or until the bacon is crisp. Serve piping hot, allowing 2 bundles per person.

———— • ————

FRENCH BEAN AND POTATO SALAD WITH MINT

1¼ lb (600 g) French beans

1¼ lb (600 g) new potatoes

juice of 1 lemon

1 sprig mint

olive oil

salt and pepper

Serve with:

stuffed tomatoes

lamb kebabs

Trim the beans, rinse, drain and boil, uncovered, for about 15 minutes or until they are tender but still crisp. Drain in a colander, refresh under running cold water and leave to cool.

Scrub the potatoes under running cold water or merely rinse if they are very clean; do not peel. Steam or boil until tender and leave to cool for an hour or more.

Peel the potatoes if wished and cut into pieces; chop the beans. Mix together in a large salad bowl or serving dish, adding salt, pepper and lemon juice to taste. Sprinkle with the apple mint leaves, torn into small pieces and serve as part of a cold summer lunch.

———•———

PAN-FRIED MANGETOUT

1¾ lb (800 g) mangetout

2 oz (60 g) butter

salt and pepper

Serve with:

hot vegetable moulds or egg dishes

Set up the steamer ready for use.

Prepare the mangetouts as described on page 22, taking care to pull off any strings along their edges. Rinse them, drain and steam for 3 minutes.

Heat the butter in a large frying pan and once the foam has subsided, stir-fry the mangetout for 3–4 minutes; they must remain crisp.

Season with a little salt and pepper and serve at once as a vegetable accompaniment to almost any dish. Mangetout go particularly well with rich, rather fatty meats such as roast duck or goose.

———•———

MANGETOUT IN CREAM AND BASIL SAUCE

1¾ lb (800 g) mangetout

1 oz (30 g) butter

7 fl oz (200 ml) crème fraîche *or single cream*

1 oz (30 g) tbsp finely grated Gruyère or Emmenthal cheese

10 basil leaves, torn into small pieces

salt and pepper

Serve with:

chicken, fish or egg dishes

Set up the steamer ready for use. Trim the mangetout (see page 22), rinse and drain them and steam for 3 minutes. Heat the butter in a large frying pan and stir-fry over a high heat for 2 minutes. Reduce the heat, add a pinch of salt and some freshly ground pepper and the cream. Cook for 5 minutes over a low heat, stirring gently to coat the vegetables with the cream. Stir in the cheese and basil and serve on hot plates. This is particularly good with egg dishes, such as omelettes, and as a crunchy contrast in texture to soufflés.

———•———

MANGETOUT WITH SHALLOTS AND BACON

1¾ lb (800 g) mangetout

4 thick rashers smoked streaky bacon

2 large shallots, peeled and finely chopped

1 sprig thyme

2 oz (50 g) butter

salt and pepper

Serve with:

Rösti Potatoes (see page 195) and omelettes, or with Cauliflower Timbale (see page 83)

Set up the steamer ready to cook the mangetout. Prepare these as described on page 22. Rinse, drain and steam for 3 minutes. Choose bacon that has a good proportion of lean to fat, remove the rind, if present, and dice the rashers. Heat a non-stick frying pan and when it is very hot, add the diced bacon and fry, stirring, until crisp and lightly browned. Take out of the pan, leaving as much fat behind as possible and drain on kitchen paper.

Heat the butter in the frying pan and fry the shallot very gently with the sprig of thyme for 10 minutes, stirring frequently. When the shallot is tender but not browned, turn up the heat, then add the mangetout and stir-fry for 4 minutes or until they are tender but still crisp. Season with a very little salt and some freshly ground pepper, stir in the crisp bacon and then take off the heat. Serve at once.

———— • ————

GREEN BEANS WITH TOMATO SAUCE

1 lb (500 g) green beans or young runner beans

1 small onion, peeled and finely chopped

1 clove garlic, peeled and crushed

lb (500 g) tinned tomatoes, drained and chopped

pinch caster sugar

pinch oregano

1 tbsp chopped parsley

1 chopped fresh basil leaves

3 tbsp olive oil

salt and pepper

Serve with:

roast chicken

grilled fish

Fill a large saucepan two-thirds full of water, add 1 tsp salt and bring to the boil. Prepare the beans as described on page 22, rinse them, drain and add to the boiling water. Boil fast, uncovered, for 10 minutes or until just tender but still very crisp. Drain in a colander and refresh under running cold water.

Heat the olive oil in a large flameproof casserole dish and fry the onion very gently with the garlic for about 10 minutes, stirring. Add the tomatoes with a pinch each of sugar and salt and some freshly ground pepper; increase the heat and bring to the boil, stirring. Simmer, uncovered, for about 10 minutes to reduce the liquid and thicken the sauce. Add the oregano, parsley and the beans and stir. Reduce the heat to low, cover and simmer gently for 4–5 minutes, or until the beans are as tender as you like them. Correct the seasoning if necessary, sprinkle with the basil and serve hot.

———— • ————

SPICED LENTILS AND SPINACH

8 oz (225 g) brown lentils

2¾ lb (1.2 kg) spinach

6 tbsp ghee (clarified butter, see page 36)

1 piece fresh ginger, peeled and finely chopped

1½ green chilli peppers, finely chopped

3 tbsp chopped coriander leaves

3 tbsp fresh lime juice
salt and pepper
Serve with:
Chapatis (see page 81)
Aubergines with Yoghurt Dressing (see page 147)

Rinse the lentils well in a sieve under running cold water. Place in a heavy-bottomed saucepan with 1½ pints (900 ml) cold water and bring slowly to the boil, skimming off any scum that rises to the surface. Reduce the heat to very low, cover, leaving a gap for the steam to escape, and simmer gently for 1 hour.

Pick over the spinach, trimming it and removing any large tough stalks (these can be used in soups and other dishes); only the leaves and tender stalks are used for this recipe. Wash very thoroughly in several changes of cold water; drain.

Heat the ghee in a large, heavy-bottomed flameproof casserole dish and stir-fry the ginger and chilli pepper briefly over a moderately low heat; stir in the coriander and add the spinach. Continue stirring until the spinach has wilted. Add the lentils, together with their cooking water, scant 1 tsp salt, some freshly ground black pepper and stir once more. Cover tightly and reduce the heat to very low; simmer for about 30 minutes, stirring occasionally, by which time the lentils should be very tender. Taste and add a little more salt and pepper if needed, mix in the lime juice and continue cooking while stirring for 2–3 minutes before removing from the heat. Serve as part of an Indian meal or as the main course for a light lunch, with a refreshing salad and chapatis.

———•———

BRAISED LENTILS

1 lb (500 g) continental lentils
4 rashers streaky bacon
1 small onion

1 medium-sized carrot
1 leek, green part only
5 fl oz (150 ml) dry white wine
8 oz (250 g) tinned tomatoes, drained and seeded
2–2¼ pints (1.2 litres) vegetable stock (see page 38) or chicken stock (see page 37)
4 tbsp olive oil
salt and pepper
Serve with:
good-quality, highly flavoured sausages

Rinse the lentils thoroughly in a sieve under running cold water. Peel the onion and the carrot; remove the tough, outermost green layers of the leek and wash the inner part thoroughly; chop these three vegetables together very finely. Heat the olive oil in a large, heavy-bottomed flameproof casserole dish and sweat the chopped vegetable mixture in it for 15 minutes, stirring frequently.

Turn up the heat, add the lentils and cook for 2–3 minutes while stirring. Pour in the wine and when it has almost completely evaporated, add the chopped tomato flesh. Pour in sufficient stock to cover the lentils. Simmer very gently with a tightly fitting lid for about 1 hour, or until the lentils are tender. Add a little salt and some freshly ground pepper; if there is a lot of liquid remaining in the casserole dish, cook, uncovered over a moderately high heat for a few minutes to reduce and thicken. Serve with full-flavoured or spicy grilled or boiled sausages or poultry dishes.

———•———

JAPANESE SWEET RED BEAN PASTE
Yude azuki

9 oz (250 g) aduki beans or red soy beans
bicarbonate of soda
10 oz (300 g) sugar

239

5 fl oz (150 ml) peanut oil
2 tbsp vanilla sugar or ½ tsp vanilla essence
Decorate with:
small mint leaves
Serve with:
vanilla ice cream
peeled, sliced kiwi fruit or fruit salad

Soak the beans overnight or for at least 12 hours in plenty of cold water; when they have finished soaking, rinse in a sieve under running cold water. Place in a large, heavy-bottomed saucepan or flameproof casserole dish with 2¾ pints (1½ litres) cold water. Bring quickly to the boil over a high heat, turn down the heat to low, add a pinch of bicarbonate of soda, stir and cover, leaving a chink for the steam to escape. Simmer gently for 1½–2 hours, until the beans are very tender. Add a small pinch of salt when they are nearly done. Remove from the heat and put the beans and any remaining liquid through the vegetable mill with a fine-gauge disc fitted.

Return the purée to the saucepan or casserole dish and stir in the sugar and oil. When these have been added to the purée, stir in the vanilla sugar or vanilla essence. Cook the purée over a low heat, stirring continuously, until it thickens to a paste. Leave to cool at room temperature before packing into sterilized glass preserving or jam jars with airtight seals or lids. This red bean paste goes very well with vanilla ice cream with kiwi fruit or a mixed fruit salad; use a small scoop to add 1 or 2 portions of the paste to each ice cream serving.

——— • ———

ADUKI JELLY
Mizuyokan

8 oz (250 g) aduki beans or red soy beans
bicarbonate of soda
preserving sugar (cane sugar)
2 tbsp vanilla sugar or few drops vanilla essence
2 tbsp (2 sachets) gelatine powder
salt
Garnish with:
small mint leaves
Serve with:
fruit salad or vanilla ice cream

Follow the directions given in the previous recipe for Japanese Sweet Red Bean Paste for soaking and cooking the aduki beans. Sweeten the purée but do not add any oil. Simmer the purée until it has reduced to a paste. In the early stages it will only need stirring occasionally but as it thickens you will need to stir almost continuously, otherwise it may catch and burn because no oil has been added.

Remove the red bean paste from the heat and pour 3½ fl oz (100 ml) water into a small, heavy-bottomed saucepan; add 1 tbsp caster sugar and the agar-agar powder and simmer gently for a few minutes while stirring until the sugar and agar-agar have completely dissolved. Stir into the hot bean paste, mixing very thoroughly. Rinse the inside of a small, rectangular dish with cold water (you will need a dish of about 1¼ pints (700 ml) capacity) and fill with the mixture. Tap the bottom of the dish gently on the work top to make it settle and smooth the surface level with a palette knife. Leave to set (agar-agar will set at normal room temperature) then chill in the refrigerator unless you wish to serve it soon after it has jelled.

To serve the jelly, run the blade of a knife around the inside of the dish or container to loosen, and turn out on to a pastry board; cut it into fairly thick slices and garnish with mint leaves.

·MUSHROOMS AND TRUFFLES·

MUSHROOM AND PARMESAN SALAD

p. 249

Preparation: 20 minutes
Difficulty: very easy
Appetizer

MUSHROOM SALAD WITH EMMENTHAL AND CELERY

p. 249

Preparation: 30 minutes
Difficulty: very easy
Appetizer

MUSHROOM AND TRUFFLE SALAD

p. 249

Preparation: 15 minutes
Difficulty: very easy
Appetizer

HOT MUSHROOM TERRINE WITH CELERIAC AND TOMATO SAUCE

p. 249

Preparation: 40 minutes
Cooking time: 50 minutes
Difficulty: easy
Appetizer

MUSHROOM AND JERUSALEM ARTICHOKE SALAD

p. 250

Preparation: 35 minutes
Difficulty: easy
Appetizer

MUSHROOM, RICE AND TRUFFLE SALAD WITH SHERRY MAYONNAISE

p. 250

Preparation: 30 minutes
Cooking time: 20 minutes
Difficulty: easy
Appetizer

MUSHROOM AND NOODLE SALAD WITH SESAME SAUCE

Rien pan san su

p. 251

Preparation: 30 minutes
Cooking time: 15 minutes
Difficulty: easy
Appetizer

HONEY MUSHROOMS WITH ASPARAGUS TIPS AND BASIL SAUCE

p. 252

Preparation: 35 minutes
Cooking time: 15 minutes
Difficulty: easy
Appetizer

Truffled eggs
with polenta

*Mushrooms Bordeaux
style*

242

*Chinese chicken and
mushroom soup*

*Gnocchi with
mushroom sauce*

DEEP-FRIED MUSHROOMS WITH TARTARE SAUCE

p. 261

*Preparation: 25 minutes
Cooking time: 15 minutes
Difficulty: easy
Main course*

MUSHROOM AND HERB OMELETTE

p. 261

*Preparation: 20 minutes
Cooking time: 15 minutes
Difficulty: easy
Main course*

POLENTA WITH MUSHROOM TOPPING

p. 262

*Preparation: 30 minutes
Cooking time: 1 hour
Difficulty: easy
Main course*

BREADED MUSHROOMS

p. 262

*Preparation: 15 minutes
Cooking time: 25 minutes
Difficulty: easy
Main course*

MUSHROOM RISSOLES WITH FRESH TOMATO SAUCE

p. 263

*Preparation: 25 minutes
Cooking time: 30 minutes
Difficulty: easy
Main course*

TRUFFLED EGGS WITH POLENTA

p. 263

*Preparation: 5 minutes
Cooking time: 40 minutes
Difficulty: easy
Main course*

MUSHROOMS BORDEAUX STYLE

p. 264

*Preparation: 25 minutes
Cooking time: 10 minutes
Difficulty: easy
Accompaniment*

MOREL MUSHROOMS AND ASPARAGUS WITH MOUSSELINE SAUCE

p. 264

*Preparation: 30 minutes
Cooking time: 12 minutes
Difficulty: easy
Accompaniment or appetizer*

248

MUSHROOM AND PARMESAN SALAD

14 oz (400 g) mushrooms (see method)
1 1½-oz (40-g) piece Parmesan cheese with no rind
olive oil
salt

Use freshly gathered young field mushrooms or the freshest button mushrooms for this recipe. Wipe the caps with a damp cloth and trim off the ends of the stalks. Just before serving, cut the mushrooms into thin slices, arrange on individual plates, sprinkle with a small pinch of salt and a little oil. Scatter flakes or small slivers of Parmesan cheese over the mushrooms and serve.

———•———

MUSHROOM SALAD WITH EMMENTHAL AND CELERY

14 oz (400 g) mushrooms (see method)
3 oz (90 g) thinly sliced Emmenthal cheese
2 green celery stalks
olive oil
salt

See preceding recipe for choice and preparation of mushrooms. Prepare them but do not slice until just before serving. Cut the cheese into thin strips; any good sliceable semi-hard type will do, but Emmenthal goes particularly well with mushrooms. Wash, dry and trim the celery stalks and run the potato peeler down the curved, outer sides to get rid of any strings. Chop finely.

Mix the cheese and celery in a bowl with a little oil and salt. Transfer to individual plates and slice the mushrooms thinly over the top. Sprinkle with a little more oil and salt and serve at once.

MUSHROOM AND TRUFFLE SALAD

14 oz (400 g) button mushrooms
1–2 tinned truffles
extra-virgin olive oil
1–1½ tsp fresh lemon juice
salt

Prepare the mushrooms as described for Mushroom and Parmesan Salad on this page. Reserve the liquid from the tin of truffles. Tinned pieces of truffle are cheaper than whole ones. Slice the mushrooms very thinly into a bowl just before serving, sprinkle with a little oil, 1–1½ tsp lemon juice, a pinch of salt and the reserved truffle liquid.

———•———

HOT MUSHROOM TERRINE WITH CELERIAC AND TOMATO SAUCE

6 oz (180 g) fresh, trimmed cleaned ceps (boletus edulis)
2 tbsp olive oil
½ clove garlic, finely chopped
2 tbsp finely chopped parsley
7 fl oz (200 ml) double cream
7 fl oz (200 ml) milk
2 eggs
4 extra egg yolks
nutmeg
1 tbsp grated hard cheese
2 tbsp plain flour
salt and pepper
For the celeriac and tomato sauce:
10 oz (300 g) celeriac
4 tbsp olive oil
2 sprigs thyme

249

2 large ripe tomatoes or 4 tinned tomatoes
4–6 large sprigs basil
1 lemon
scant 1 oz (20 g) unsalted butter
salt and pepper

Preheat the oven to 350°F (180°C) mark 4.

Prepare the mushrooms by wiping with a damp cloth and trimming off the ends of their stalks; slice thinly and cook in the oil with the garlic and half the chopped parsley, stirring to cook lightly and evenly; make sure that the moisture they release evaporates. Season with a little salt and pepper; process one third of the mushroom mixture in the blender with the cream, milk, 2 whole eggs and the extra yolks, the Parmesan and a pinch each of salt, pepper and nutmeg. When smoothly blended, transfer the mixture into a bowl and mix in the reserved mushrooms and the rest of the parsley evenly.

Grease a 2-pint (1.2-litre) lidded terrine dish with butter and line the bottom with greased waxed paper. Fill with the mixture, cover with the lid, sealing the join with a luting paste made with 2 tbsp flour mixed with a little cold water; leave the small hole in the lid free for the steam to escape. Alternatively, cover tightly with a sheet of foil in which you have pricked a few small holes. Place in a roasting tin, add sufficient boiling water to come half way up the sides of the terrine and place in the oven. Cook for 50 minutes.

Make the sauce. Peel the celeriac, slice very thinly using a mandoline cutter or food processor; cut these slices into julienne strips (see page 24). Heat the oil in a wide, fairly deep frying pan with the thyme and stir-fry the celeriac over a moderately high heat until tender but still fairly crisp. Add the peeled, seeded and coarsely chopped tomatoes and the basil leaves cut into thin strips. Season with salt and pepper and simmer for 2–3 minutes while stirring; set aside until the mushroom terrine is cooked, then reheat the sauce just before serving, adding 1 tsp lemon juice. Stir in the butter off the heat. Turn out the hot, cooked terrine and serve with the sauce. Serves 10–12.

MUSHROOM AND JERUSALEM ARTICHOKE SALAD

7 oz (200 g) fresh ceps (boletus edulis) *or substitute (see method)*
10 oz (300 g) Jerusalem artichokes
1 lemon
olive oil
salt and pepper
Garnish with:
parsley sprigs
radish flowers (see page 28)

Substitute large, closed cap cultivated mushrooms or freshly gathered field mushrooms for the ceps if necessary. Wipe with a damp cloth and trim off the ends of the stalks. If preparing several hours in advance, wrap in a damp cloth and chill in the salad compartment of the refrigerator. Have a large bowl of cold water mixed with the freshly pressed lemon juice; peel the artichokes, dropping each one into the acidulated water as soon as you have finished peeling it, to prevent discoloration.

Shortly before serving the salad, slice the mushroom caps and stalks thinly. Cut the well drained and dried Jerusalem artichokes into very thin slices. Mix the vegetables gently in a large salad bowl, adding 1–2 tbsp oil, salt and pepper.

———— • ————

MUSHROOM, RICE AND TRUFFLE SALAD WITH SHERRY MAYONNAISE

1 fresh or tinned white or black truffle
14 oz (400 g) risotto rice
7 fl oz (200 ml) dry white wine

4 oz (120 g) fresh asparagus tips
4 oz (120 g) tender white celery stalks
4 oz (120 g) ceps (boletus edulis) or closed cap cultivated mushrooms (net weight)
For the sherry mayonnaise:
yolk from 1 large, hard-boiled egg
2 tbsp lemon juice
2 tbsp dry sherry
1 tsp mild French mustard
½ tsp paprika
7 tbsp light olive oil
3 tbsp single or double cream
salt and freshly ground white pepper

Preheat the oven to 350°F (170°C) mark 3. Sprinkle the rice into a pan of fast boiling salted water; boil fast for 8 minutes, then drain, transfer to a casserole dish greased with butter and sprinkle the wine all over it, together with the juice from tinned truffles if used. Cover tightly and cook in the oven until it has absorbed all the wine and is tender. Take out of the oven, remove the lid and leave to cool. Steam or boil the well washed asparagus tips until just tender. Cut the celery into strips.

Make the sherry mayonnaise: crush the hard-boiled egg yolk in a bowl with a fork, gradually work in the lemon juice, sherry, mustard, paprika and a little salt and pepper. Beat with the fork or a small balloon whisk as you very gradually add the oil, followed by the cream, adding this a little at a time. Combine the rice gently but thoroughly with the celery, asparagus tips and thinly sliced mushrooms, together with the mayonnaise. Slice the truffle in wafer-thin pieces all over the surface, and serve immediately.

———— • ————

MUSHROOM AND NOODLE SALAD WITH SESAME SAUCE
Rien pan san su

4 dried mushroom caps (shitake, or Chinese dried mushrooms)
1 packet cellophane noodles
2 eggs
pinch caster sugar
pinch ajinomoto (Japanese taste powder) (optional)
2 thick slices lean cooked ham, cut in strips
1 cucumber
sunflower or peanut oil
salt
For the sesame sauce:
4 fl oz (120 ml) sesame oil
2½ fl oz (70 ml) sweet Japanese vinegar
1 tsp Japanese wasabi powder
3 tbsp Japanese (light) soy sauce
3 tbsp mirin or sweet saké

small pinch ajinomoto (Japanese taste powder) (optional)

Cook the Chinese cellophane noodles (a packet usually weighs about 8–9 oz/225–250 g). Soak these in a bowl of hot water for 30 minutes. Put the dried mushrooms to soak in hot water for 30 minutes, then drain off the water, reserving 5 fl oz (150 ml) of it and squeeze excess moisture out of the mushrooms. Cut the caps into strips. Beat the eggs with a pinch each of salt and caster sugar and a small pinch of *ajinomoto* taste powder (optional). Heat 1 tbsp sunflower oil in a wide non-stick frying pan and add the beaten egg mixture, tipping the pan this way and that to spread it out thinly all over the bottom of the pan. Cook over a low heat for a few minutes until completely set but not at all browned. Slide the very thin omelette out of the pan on to a plate and cut into strips.

Rinse and dry the cucumber; cut off both ends. Rub the entire surface with plenty of salt, leave for a few minutes and then wipe very thoroughly with kitchen paper or a clean cloth. Slice into strips. Make the sauce by mixing all the ingredients thoroughly in a bowl, diluting with the strained reserved water from the soaked mushrooms.

Add the cellophane noodles to the boiling water, cook for 3 minutes only, then drain in a sieve. Cut into short lengths, transfer to a serving bowl, and arrange the ham, cucumber, mushroom and omelette strips all over the surface. Cover and chill. Serve cold with the sesame sauce.

———•———

HONEY MUSHROOMS WITH ASPARAGUS TIPS AND BASIL SAUCE

14 oz (400 g) honey mushrooms or substitute (see method)

16 large, fat asparagus spears

1 large shallot, finely chopped

1 clove garlic, crushed

3½ fl oz (100 ml) light stock

3 basil leaves

3 tbsp light olive oil

salt and pepper

For the basil sauce:

4 tbsp pesto *sauce (see page 177)*

3 tbsp single cream

salt

Prepare the asparagus spears as directed on page 17; double the number of spears if you are using thin, green asparagus. Cut off about 2 in (5 cm) from the larger, tougher ends of the stalks. Make up 4 equal bundles and tie each securely but not too tightly with two pieces of string; steam or boil in salted water for about 12 minutes. If they are ready before you have finished preparing and cooking the mushrooms, drain and keep hot.

If you prefer not to gather wild honey mushrooms

(also known as bootlace fungus) use very fresh baby button mushrooms. If using honey mushrooms, separate the caps from the stalks, cut off the tough ends of the stalks; rinse and drain the caps and stalks well. Cut the larger stalks lengthwise in half. Heat the oil in a very wide non-stick frying pan and stir-fry the mushrooms with the shallot and the garlic over a high heat for 2–3 minutes. Season with a little salt and pepper and add the stock and the coarsely snipped basil leaves. Cover and simmer gently for 6 minutes or until the mushrooms are cooked. Stir the cream into the *pesto*. Untie the asparagus bundles and arrange the spears on each individual plate, spoon a little creamy pesto sauce over the tips and place the mushrooms to one side of them.

———•———

CHICKEN AND TRUFFLE SALAD WITH ANCHOVY DRESSING

1 black truffle (tinned in juice or preserved in oil)

7 oz (200 g) mixed salad leaves

2 boneless skinned chicken breasts, each weighing 4½–5 oz (140 g)

1 tbsp fresh lemon juice

3½ fl oz (200 ml) chicken stock (see page 37)

3½ fl oz (200 ml) dry white wine

2 finely chopped shallots

2 sprigs thyme

salt and pepper

For the anchovy dressing:

2 tsp juice or oil from tinned or bottled truffle

2 tbsp lemon juice

2 tbsp white wine vinegar

4 fl oz (120 ml) olive oil

2 small tinned anchovy fillets

1 tbsp chopped chives

pepper

Season the chicken breasts lightly with salt and pepper and sprinkle with lemon juice on both sides. Arrange a bed of salad leaves on each plate. Place the chicken breasts and any unabsorbed juice in a fairly small non-stick saucepan with the stock, wine, shallot and thyme. Heat to just below boiling point, cover, reduce the heat further and barely simmer for 6 minutes, turning after 3 minutes. If the liquid boils, the chicken will be tough. Test by inserting a knife into the thickest part; if the juice runs out at all pink it is not cooked, continue the simmering process until done.

Make the anchovy dressing by placing all the ingredients in the blender and processing at high speed until smoothly blended. Take the chicken breasts out of the liquid; place on the chopping board and cut across into thin slices. Fan some of these slices out, slightly overlapping, on each bed of salad greens, cover them with a thin coating of anchovy sauce and sprinkle with wafer-thin slivers of truffle. Serve while the chicken is still slightly warm.

———•———

MUSHROOM SOUP

½ lb (250 g) ox-tongue or beefsteak fungus or field or horse mushrooms (see method) (net trimmed weight)
1½ pints (900 ml) chicken stock (see page 37)
2 shallots, finely chopped
1 clove garlic, crushed
1 very large tomato, peeled, seeded and diced
1 tbsp chopped parsley
4 thick slices coarse white bread
2 oz (50 g) grated Parmesan cheese
8 basil leaves
olive oil
salt and freshly ground pepper

Heat the stock while you trim the mushrooms. You may prefer to use brown or white cultivated mushrooms for this recipe. Cut into fairly small dice. Heat

3 tbsp olive oil in a wide, shallow saucepan and fry the shallots and garlic gently for 5 minutes, stirring frequently. Turn up the heat, add the diced mushrooms and a pinch of salt and stir-fry for about 2 minutes. Add the boiling stock and the diced tomato flesh. Simmer, uncovered, for about 8 minutes or until the mushrooms are cooked. Sprinkle in the parsley.

Heat the bread slices in a low oven until crisp and dry. Place 1 slice in each soup bowl; sprinkle each slice with some of the Parmesan cheese and a very little olive oil, grind plenty of black pepper on to each slice and ladle in the mushroom soup. Decorate with the fresh basil leaves.

———•———

CHINESE CHICKEN AND MUSHROOM SOUP

½ oz (15 g) dried Chinese mushroom caps (shitake)
1 oz (20 g) black frilly Chinese fungus (Jew's ears or tree ears)
2½ oz (60 g) raw chicken breast, cut into strips
2–2½ oz (60 g) fillets of very fresh white fish (e.g. Dover sole, haddock)
2–2½ oz (60 g) peeled prawns
1¾ pints (1 litre) chicken, shellfish or vegetable stock (see pages 37 and 38)
1 oz (30 g) tinned bamboo shoots, drained
2 oz (50 g) trimmed spinach leaves, cut lengthwise in half
salt
For the marinade:
pinch monosodium glutamate (optional)
pinch freshly ground white pepper
1 tbsp Chinese rice wine or dry sherry
1 tbsp cornflour
½ tsp salt

Soak both sorts of Chinese mushrooms separately in warm water for at least 30 minutes. Cut off and discard the stalks from the shitake and trim off any tough pieces from the frilly black fungus. Rinse and squeeze out excess moisture. Cut into strips and set aside. Mix the marinade ingredients in a large bowl. Add the chicken, fish and prawns and mix thoroughly. Leave to stand for 15 minutes.

Bring the stock to the boil and stir in all the ingredients at once except for the spinach. Simmer gently for 3 minutes and then add the well washed spinach. Simmer for another minute, then serve.

——•——

MUSHROOM RISOTTO

5–6 oz (160 g) fresh ceps (boletus edulis) or 2–3 oz (50–90 g) dried mushrooms
3 tbsp clarified butter (see page 36)
½ small clove garlic, crushed (optional)
1 tbsp finely chopped parsley (optional)
For the basic risotto:
11½ oz (320 g) risotto rice (e.g. arborio)
2¾ pints (1½ litres) chicken stock (see page 37)
2½ oz (70 g) butter
2 shallots, very finely chopped
2½ fl oz (70 ml) dry vermouth or dry white wine
4 tbsp grated Parmesan cheese
salt and freshly ground white pepper

If you are using dried mushrooms, put them to soak in a bowl of warm water 30 minutes before starting preparation. Set the stock to come to boiling point over a moderate heat. While it is heating, melt 1 oz (30 g) of the butter in a fairly deep, wide heavy-bottomed frying pan or a large, shallow saucepan and cook the shallot over a low heat for 5 minutes, stirring frequently. Add the rice and cook, stirring over a slightly higher heat for 1½ minutes. Sprinkle in the dry vermouth, cook briefly, until it has evaporated and then add about 8 fl oz (225 ml) of boiling hot stock. Cook until the rice has absorbed almost all the liquid, stirring from time to time, then add more stock, and continue cooking with the occasional stir, adding more boiling liquid whenever necessary. Test the rice after 14 minutes; it should be tender but firm. Prepare the mushrooms. Wipe with a slightly dampened cloth and cut off the stalks. Rinse briefly if wished, dry immediately with kitchen paper and slice. If you are using dried mushrooms, drain them, squeeze out excess moisture and cut into small pieces. Heat the clarified butter and stir-fry the mushrooms for 2–3 minutes at most over a fairly high heat. Season lightly with salt and freshly ground pepper. As soon as the risotto is ready, remove from the heat, add the remaining, solid, butter, the Parmesan cheese and the mushrooms; fold into the risotto. Add the garlic and parsley.

——•——

FRIED RICE WITH MUSHROOMS

8 oz (225 g) hedgehog fungus or substitute (see method)
10 oz (300 g) cold cooked rice
3 large cloves garlic, crushed
2–3 sage leaves
1 oz (20 g) butter
3 tbsp light olive oil
1 tbsp finely chopped parsley
salt and pepper

Cook the rice until tender the day before you plan to serve this dish and chill in the refrigerator in an airtight container; take it out of the refrigerator 30 minutes before cooking to bring to room temperature. Trim, briefly rinse and dry the mushrooms with kitchen paper. If you cannot gather your own wild fungus (this is one of the safest wild mushrooms to gather as it is extremely difficult to confuse with any of the inedible fungi) use very fresh closed cap brown or white button mushrooms. Cut fresh hedgehog mushrooms or the larger cultivated variety in half and then slice; slice the trimmed stalks. Heat the oil and butter for 30 seconds in a wok or large frying pan with the garlic and sage leaves. Turn up the heat to high, add the rice and 1 tsp salt and stir-fry for 2 minutes. Reduce the heat and break up any rice that has stuck together; continue doing this for 4–5 minutes while the rice heats through. Season to taste.

Place the sliced mushrooms on top of the rice and sprinkle with parsley.

———•———

TRUFFLE RISOTTO

1 small white or black fresh or tinned truffle

For the risotto:

1 quantity Mushroom Risotto (see page 254); use an extra 3 tbsp grated Parmesan and omit the mushrooms and garlic

Make the basic risotto. If using fresh truffles brush with a soft brush and place in the centre of the table with a truffle slicer for each person to slice some wafer-thin shavings of the raw truffle over their risotto. Alternatively, use tinned whole black truffles and slice very thinly. Stir the truffle into the finished risotto at the same time as you add the grated Parmesan cheese and remaining, solid butter.

———•———

PASTA WITH MUSHROOMS

7 oz (200 g) fresh ceps (boletus edulis) (net prepared weight) or 2–3 oz (60–90 g) dried ceps

8–9 oz (250 g) fresh green ribbon noodles

2 cloves garlic, crushed

3 tbsp olive oil

2 tbsp chopped parsley

½ oz (15 g) butter

salt and pepper

If dried mushrooms are used, soak them for at least 30 minutes in warm water. Set a large saucepan of salted water to come to the boil. Trim the fresh mushrooms and slice about ⅛ in (3 mm) thick. Heat the oil in a wide, fairly shallow saucepan or deep frying pan and fry the mushrooms with the garlic over a moderate heat for 2 minutes. If dried mushrooms are used, drain and squeeze out excess moisture, then blot dry on kitchen paper before frying. Lower the heat; add the parsley, a little salt and pepper and 2 tbsp of the boiling salted water. Cover and simmer for 3–4 minutes. Draw aside from the heat, add a little salt to taste and stir in the remaining, solid butter until it has just melted.

Add the pasta to the fast boiling water; the noodles will only take a very few minutes to cook if fresh. Drain and transfer to heated plates. Top each serving with the mushrooms. Serve while still very hot.

———•———

TAGLIATELLE WITH TRUFFLE

1 small white or black truffle

8–9 oz (250 g) fresh ribbon noodles

2 oz (50 g) butter

1½ oz (50 g) grated Parmesan cheese

salt

Bring a large saucepan of salted water to a full boil, add the pasta and cook for 2–3 minutes only if fresh. Commercially prepared dry pasta will take a little longer and you will need only about ½–⅔ of this weight for a first course; drain when tender but still firm. Drain briefly, leaving 2–3 tbsp of water behind in the pan with the pasta. Stir in the solid butter and cheese over low heat. Transfer to individual heated bowls and serve with wafer-thin slices of fresh white truffle.

PASTA WITH MUSHROOMS AND PINE NUTS

1 oz (30 g) dried ceps (boletus edulis*)*
8–9 oz (250 g) fresh broad ribbon noodles
1 baby carrot
½ small onion
½ celery stalk
1 clove garlic
2 tbsp chopped parsley
3 tbsp olive oil
1 bay leaf
2 tbsp finely chopped tomato
5 fl oz (150 ml) chicken stock (see page 37)
1½ tbsp pine nuts
2 oz (50 g) grated Parmesan cheese
salt and pepper

Soak the dried mushrooms for at least 30 minutes in warm water. Heat plenty of salted water in a large saucepan. Drain the mushrooms. Squeeze out excess moisture and chop coarsely. Heat the olive oil and sweat the mixed very finely chopped carrot, onion, celery, garlic, parsley and bay leaf in it over a very low heat for 10 minutes with the lid on, stirring occasionally. Turn up the heat, add the mushrooms and a little salt and pepper; stir while cooking for 2–3 minutes. Add the chopped fresh tomato and the stock and simmer for 20 minutes, positioning the lid to leave a gap so the sauce can reduce and thicken gradually. After this time has passed, remove the lid and cook over a slightly higher heat to thicken further. Cook the pine nuts with a very little olive oil in a non-stick frying pan for 1–2 minutes, stirring, until they are pale golden brown. Set aside to add to the sauce at the last minute.

Add the noodles to the fast boiling water and cook until just tender; drain and turn into a very hot serving dish. Stir the pine nuts into the sauce, add this to the pasta and stir. Sprinkle with Parmesan cheese and serve.

MUSHROOM VOL-AU-VENTS POLISH STYLE

14 oz (400 g) fresh ceps (boletus edulis*) or ox tongue or beefsteak fungus*
8 vol-au-vent cases
3½ oz (100 g) finely chopped onion
1 shallot, finely chopped
1½ oz (40 g) butter
2 fl oz (60 ml) dry sherry or dry white wine
2 fl oz (60 ml) chicken stock (see page 37)
1½ tbsp paprika
pinch cayenne pepper or chilli powder
1 tbsp plain flour
3½ fl oz (100 ml) milk
3½ fl oz (100 ml) soured cream
1½ tbsp finely chopped parsley
salt and pepper

Preheat the oven to 350°F (180°C) mark 4; heat the vol-au-vent cases for 10 minutes before you fill them. Make the filling: trim and wipe the mushrooms with a damp cloth or rinse briefly and dry with kitchen paper. Chop coarsely. If you use dried *boletus edulis* mushrooms, allow about 4 oz (120 g) and soak for at least 30 minutes in lukewarm water before draining and squeezing out excess moisture. Chop fresh or dried, presoaked mushrooms coarsely. Cook the onion and shallot very gently in a very wide saucepan in the butter for 10 minutes, stirring frequently. When wilted and tender add the mushrooms and sauté over a higher heat for 2–3 minutes. Season with a little salt and pepper and add the sherry and stock. Stir, cooking for 2–3 minutes, to allow the liquid to reduce. Reduce the heat a little, sprinkle in the paprika, cayenne or chilli powder and the flour, stir well and then gradually stir in the milk and cream. Cover and cook over low heat for 4 minutes, then remove the lid and reduce the sauce over a moderate heat for a few minutes. Add the parsley. Fill the hot vol-au-vent cases with this mixture and serve.

GNOCCHI WITH MUSHROOM SAUCE

1¼ lb (600 g) mixed wild and cultivated mushrooms
1 packet plain potato gnocchi (see method)
1 medium-sized onion, finely chopped
2 sage leaves
1 tbsp meat extract
1 small tin tomatoes
½ small clove garlic, crushed
1½ tbsp finely chopped parsley
5 tbsp grated Parmesan cheese
½ oz (15 g) butter
salt and pepper

If you prefer to make your own *gnocchi*, follow the recipe on page 177. Trim the mushrooms.
Cook the onion and sage leaves in the oil over low heat for 10 minutes, in a fairly deep, very wide frying pan or large, shallow saucepan. Stir frequently. Turn up the heat and add the mushrooms. Sprinkle with a pinch of salt and a generous amount of freshly ground pepper and sauté for 2–3 minutes. Dilute the

rich gravy with 2 tbsp boiling water and add to the mushrooms, together with the tinned tomatoes. Stir, reduce the heat, cover and simmer gently for about 10 minutes or until the mushrooms are tender, stirring occasionally. Keep hot.
A few minutes before it is time to serve this dish, add the potato gnocchi to a very large saucepan of fast boiling salted water and as soon as they bob up to the surface, drain well. While they are cooking, take the hot sauce off the heat, stir in the garlic, parsley, Parmesan cheese, and the solid butter; as soon as the butter has melted, pour the sauce over the gnocchi and serve at once.

STUFFED MUSHROOM AND POTATO BAKE

8 large, open cap mushrooms
4 medium-sized, waxy potatoes
3 tbsp finely chopped parsley
1–2 cloves garlic, peeled and finely chopped
olive oil
oregano
salt and freshly ground pepper

Preheat the oven to 350°F (180°C) mark 4. Grease a gratin dish. Peel and slice the potatoes. Spread these out in a single layer in the dish, season and add 7 fl oz (200 ml) water. Trim the mushrooms, cutting the stalks off. Chop the stalks finely with the parsley and garlic and then mix in a bowl with 1½ tbsp olive oil and a pinch each of oregano, salt and pepper.
Spread the mixture neatly over each one; spread out in a single layer on top of the potatoes and cover with greased foil. Bake for 40–45 minutes.

MUSHROOM AND ASPARAGUS PIE WITH FRESH TOMATO SAUCE

5 oz (150 g) fresh ceps (boletus edulis)*, wild field mushrooms or large closed or open cap cultivated mushrooms*

18 large asparagus spears

2–2½ oz (60 g) celeriac

1 large shallot

1 sprig thyme

1 tbsp chopped basil

3 oz (80 g) Emmenthal cheese

3 oz (80 g) smoked cheese

8–9 oz (250 g) packed frozen puff pastry, thawed

1 egg

1 tbsp poppy seeds

butter

2 tbsp olive oil

salt and pepper

For the fresh tomato sauce:

1 lb (500 g) ripe tomatoes

1–1½ tbsp capers

1 small anchovy fillet

½ clove garlic

generous pinch oregano

4–5 basil leaves, finely chopped

1 lemon

2 tbsp extra-virgin olive oil

salt and freshly ground pepper

Preheat the oven to 400°F (200°C) mark 6. Trim and wipe or rinse and dry the mushrooms. If dried mushrooms are used, allow 1½–2 oz (40–60 g), soak in warm water for at least 30 minutes then drain, squeeze out excess moisture and blot dry. Prepare the asparagus (see page 17) and steam or boil until they are just tender. Peel the celeriac, grate it fairly coarsely and immediately sauté it in a large non-stick frying pan or saucepan over a moderate heat in a little butter, until it is wilted. Season and set aside. Cook the finely chopped shallot in 2 tbsp olive oil in the non-stick frying pan with the mushrooms and thyme until just tender. Add the chopped basil and season lightly with salt and pepper. Stir and remove from the heat. Grate the two types of cheese, keeping them separate. Roll out the pastry into a fairly thin, square sheet on a lightly floured board or working surface. Cut the square in half and place one half on a greased baking tray or in a Swiss roll tin. Beat the egg lightly in a small bowl or cup and brush half of it all over the surface of this first pastry sheet, using a pastry brush, using about half the egg and stopping short of the edges. Spread out the mushroom mixture evenly over the glazed surface, stopping ½ in (just over 1 cm) short of the edges all the way round. Arrange half the asparagus spears head to tail, spacing them out evenly in a single layer on top of the mushroom mixture. Cover them evenly with the celeriac. Top with the Emmenthal. Arrange another layer of the remaining asparagus and top with the grated smoked cheese. Cover with the other half of the pastry and seal the edges well by pressing them together all the way round.

Brush the remaining beaten egg all over the surface of the pie and sprinkle evenly with the poppy seeds. Bake for 20 minutes.

Make the sauce. Blanch, peel and seed the tomatoes and place them in the blender. Add the capers, anchovy, garlic, oregano, 2 tsp extra-virgin olive oil, salt and freshly ground pepper and process at high speed until smooth. Stir in the chopped basil and a little lemon juice. Serve with the hot pie.

———— · ————

MUSHROOM AND SPINACH STRUDEL WITH RED WINE SAUCE

7 oz (200 g) ceps (boletus edulis) *or cultivated mushrooms*

1–1¼ lb (500 g) spinach (unprepared weight)

scant 1 oz (20 g) butter

2 shallots, finely chopped
2 tbsp oil
1 clove garlic
1½ tbsp finely chopped parsley
7 oz (200 g) ricotta cheese
2 eggs
2 oz (50 g) freshly grated Parmesan cheese
8–9 oz (250 g) packet frozen puff pastry, thawed
1 tbsp white sesame seeds
nutmeg
salt and pepper
For the red wine sauce:
3½ oz (100 g) ham
3½ oz (100 ml) dry red wine
1½ tbsp marsala
1 shallot, finely chopped
1 sprig fresh thyme
1 bay leaf
9 fl oz (250 ml) beef stock (see page 37)
½ tsp cornflour or potato flour
black pepper

Heat plenty of water in a large saucepan, adding a generous pinch of salt. Preheat the oven to 400°F (200°C) mark 6. Trim the mushrooms, slice thinly and sauté in the oil with a whole, unpeeled clove of garlic. Sprinkle with salt and pepper, stir in the parsley and set aside. Trim and wash the spinach very thoroughly; use the leaves only for this recipe. Add to the large pan of boiling salted water and blanch for 3 minutes; drain well. Heat the butter in a frying pan, fry the shallot gently until tender, add the spinach and sauté over a higher heat for 2–3 minutes. Add a pinch of freshly grated nutmeg and set aside. Beat the ricotta with a wooden spoon briefly, then beat in 1 whole egg and the Parmesan cheese, a pinch of nutmeg, salt and pepper and a very little more garlic if wished, finely crushed.

Roll out the pastry into a large square, cut in half and place one half on a baking tray or sheet. Spread all the spinach over the pastry, keeping well clear of the edges; cover with the mushroom mixture and then top with the ricotta, using a piping bag with a large, plain nozzle to make this easier. Cover with the other half of the pastry, sealing the edges very securely by pressing them together with the prongs of a fork. Beat the remaining egg lightly and brush over the surface of the pastry; sprinkle the surface with the sesame seeds. Bake for about 20 minutes, or until the pastry is crisp, and golden brown.

Use this baking time to make the sauce. Cut half the ham (2 oz/50 g) into strips. Place the rest of the ham in the blender or food processor and reduce to a smooth purée. Pour the wine into a small saucepan with the marsala, the very finely chopped shallot, sprig of thyme, bay leaf and 3–4 black peppercorns. Boil gently, uncovered, to reduce by one third. Mix the potato flour well with the cold stock and stir into the wine mixture followed by the puréed ham. Stir continuously as the mixture heats and once it has reached boiling point, simmer gently, uncovered, to reduce for about 10 minutes. Strain this rather thin sauce into another saucepan and keep warm. When the cheese and spinach pie is done, take it out of the oven. Reheat the sauce if necessary and stir in the reserved ham strips. Cut the pie across into thick slices and serve each portion with a little of the sauce.

———— • ————

MUSHROOM AND CHEESE PUFF PASTRY PIE

1½ lb (700 g) ceps (boletus edulis) *or your own choice of mushrooms*
1 clove garlic, finely chopped
2 tbsp finely chopped parsley
3 tbsp olive oil
1 8–9 oz (250 g) packet frozen puff pastry, thawed
butter
plain flour

18 fl oz (500 ml) thick béchamel sauce (see page 36)
7 oz (200 g) mild semi-soft cheese, very thinly sliced
1 egg, lightly beaten
1½ tbsp sesame seeds
salt and pepper

Preheat the oven to 400°F (200°C) mark 6. Trim and wipe the mushrooms. Cut into thin slices. Heat the oil in a wok or large frying pan and stir-fry the mushrooms with the garlic and parsley over moderately high heat for a maximum of 5 minutes, adding a pinch each of salt and freshly ground pepper when they are nearly done. Simmer briefly to reduce any juices they have produced, then set the wok aside. Make a béchamel sauce and stir in the cheese.

Cut the thawed pastry in half; roll both halves out on a lightly floured pastry board into thin discs or circles about 10½–11 in (27–28 cm) in diameter and use one to line a prepared 9½-in (24-cm) shallow tart tin or quiche dish greased with butter and lightly dusted with flour. Lightly beat the egg and use just over half to glaze this pastry case with a thin coating, using a pastry brush. Cover evenly with two thirds of the cheese sauce, spread the mushrooms out on top and then cover them with the remaining sauce. Place the other pastry disc on top as a lid, pinching the edges tightly together or pressing with the prongs of a fork to seal. Brush the remaining beaten egg over the surface of the lid and make a small, neat hole in the centre to allow the steam to escape as it cooks. Finally, sprinkle with the sesame seeds and bake for 20–25 minutes. Serves 6.

— • —

HOT MUSHROOM MOULDS WITH FRESH TOMATO SAUCE

4 oz (120 g) ceps (boletus edulis) or mushrooms of your choice (prepared weight)
1 finely chopped shallot

1 sprig thyme
3½ fl oz (100 ml) vegetable stock (see page 38) or chicken stock (see page 37)
scant 1 oz (20 g) butter
4 eggs
1½ oz (40 g) freshly grated Parmesan cheese
5 fl oz (150 ml) double cream
5 fl oz (150 ml) milk
5 fl oz (150 ml) fresh tomato sauce (see page 258)
salt and pepper

Preheat the oven to 350°F (180°C) mark 4. Trim and wipe the mushrooms and slice thinly. Sweat the shallot in the butter with the sprig of thyme for 5 minutes, increase the heat, add the mushrooms with a pinch of salt and sauté for 2–3 minutes. Add the stock, cover and simmer gently for 15 minutes (reduce this to about 5–8 minutes if using field or cultivated button mushrooms). When the mushrooms are very tender, add a little more salt if wished and plenty of freshly ground white or black pepper, remove the thyme and push the mushrooms through a fine sieve, collecting the purée in a bowl (or use a food processor). Blend in the cream and milk. Separate 2 of the eggs and beat their yolks lightly with the remaining 2 whole eggs, a pinch of salt and pepper and the Parmesan cheese. Blend into the mushroom mixture.

Grease four 7-fl oz (200-ml) capacity timbale moulds or ramekin dishes with butter. Fill them with the mushroom mixture, place well spaced out in a roasting tin or similar receptacle and pour in enough boiling water straight from the kettle to come about half to two thirds of the way up the sides of the moulds. Cook in the oven for 30 minutes; their contents will gradually thicken and set in a firm custard mixture. Make the sauce while the moulds are cooking. When they are done, turn off the oven and leave them inside for 10 minutes, still in their *bain-marie* with the door ajar. Unmould on to heated individual plates and serve with the sauce.

— • —

Deep-Fried Mushrooms with Tartare Sauce

1–1¼ lb (500 g) field mushrooms or small, open cap cultivated mushrooms
3 eggs
approx. 4 oz (120 g) fine fresh breadcrumbs
sunflower oil for frying
salt and pepper
For the tartare sauce:
4 hard-boiled eggs
1 tbsp mild mustard
8–9 fl oz (250 ml) olive oil
strained fresh lemon juice
3 gherkins, finely chopped
2–3 tbsp capers, finely chopped
2 tbsp chopped chives
2 tbsp chopped parsley
generous pinch salt
Garnish with:
1 tomato rosebud (see page 27)
small sprigs of parsley
Serve with:
steamed asparagus or artichoke hearts

Make the sauce. Take the yolks out of the hard-boiled eggs and mash very thoroughly with a fork in a bowl; work in the mustard and a generous pinch of salt. Add the olive oil very gradually, as for a classic mayonnaise, starting off by beating in a few drops at a time with a balloon whisk, gradually increasing this to a trickle. When the mixture is thick and light, stir in lemon juice to taste. Stir the chopped gherkins and capers into the sauce with the herbs. Beat in the hard-boiled egg now if wished, having first strained it through a fine sieve. Add a little more salt to taste, and season with freshly ground white pepper.

Cut the stalks off the mushrooms flush with the underside of the caps and weigh the caps. Wipe them with a slightly dampened cloth. Leave whole. Beat the eggs lightly in a bowl with a pinch of salt and some freshly ground pepper. Heat the oil in the deep-fryer to 350°F (180°C). Dip the mushrooms into this, drain briefly and then roll in the breadcrumbs, coating them all over. Deep-fry a few at a time for 2–3 minutes, until the breadcrumb coating is crisp and golden brown. Drain each batch well and keep hot, uncovered, spread out on kitchen paper while you fry the rest.

Serve as quickly as possible, with the tartare sauce.

—— • ——

Mushroom and Herb Omelette

14 oz (400 g) hedgehog fungus or chanterelles (see method)
2 shallots or ½ mild onion, finely chopped
1 oz (30 g) butter
1 tbsp oil
1 tbsp chopped summer savory leaves
1 tbsp chopped parsley
6 very fresh eggs
1½ oz (40 g) freshly grated Parmesan cheese
salt and pepper
Garnish with:
sprigs of summer savory
Serve with:
tomato salad

If you cannot obtain chanterelles, cultivated mushrooms will be a perfectly acceptable substitute for this omelette. Heat the butter and oil in a very wide, preferably non-stick frying pan and cook the shallot over a very low heat until soft, stirring frequently. Turn up the heat, add the mushrooms and sauté for about 6 minutes or until they are tender and most of the moisture they release when cooked has evaporated. Just before they are done, season with a pinch of salt and some freshly ground pepper and stir in the herbs.

261

Beat the eggs lightly in a bowl with a pinch of salt and pepper; beat in the Parmesan cheese, for which you can substitute another finely grated hard cheese if wished, provided it is not too strong. Pour this mixture into the frying pan, tipping it this way and that to ensure that the mixture spreads out evenly all over the bottom.

Run a palette knife round the edges and tip the uncooked, runny egg towards the sides; fold over before sliding out of the pan. Serve at once.

———•———

POLENTA WITH MUSHROOM TOPPING

1–1¼ lb (500 g) mixed edible, wild and cultivated mushrooms

2 large shallots or ½ mild onion, finely chopped

4 tbsp light olive oil

2 tbsp finely chopped parsley

1 tbsp plain flour

7 fl oz (200 ml) vegetable stock (see page 38) or chicken stock (see page 37)

2 tbsp grated Parmesan or other hard cheese

1 oz (30 g) butter

salt and pepper

For the polenta:

1 lb 2 oz (500 g) polenta or cornmeal

scant 3¼ pints (1.8 litres) water

salt

Make the polenta: heat the water to boiling with 1 tbsp salt in the top of a double boiler over a moderate heat and then place this over the bottom of the double boiler containing boiling water before adding the cornmeal. Sprinkle in the cornmeal or pour in a small, steady stream, stirring continu-

ously. Keep stirring as the mixture thickens. Cover and keep hot over the hot water.

Make the mushroom topping: Trim the mushrooms. Separate the stalks and cut into thick slices. Heat the oil in a very large, fairly deep frying pan and cook the shallot gently for 5 minutes. Stir in the parsley and cook for a few seconds. Turn up the heat and add the mushrooms and a pinch of salt.

Sauté the mushrooms over a high heat for 2–3 minutes, sprinkle with the flour, stir well and continue stirring as you add the stock. Bring to the boil. Reduce the heat to very low, cover and simmer gently for 5–10 minutes. When the mushrooms are tender, remove from the heat and stir in the solid butter and the Parmesan cheese until the butter has melted. Serve the piping hot polenta on heated plates, spooning a topping of the mushrooms on to each serving.

———•———

BREADED MUSHROOMS

4 very large horse mushrooms (wild or cultivated) or wild parasol mushrooms

plain flour

2 eggs

breadcrumbs

butter

olive oil

salt and pepper

Garnish with:

lemon wedges

sprigs of parsley

Beat the eggs lightly in a bowl with a pinch of salt and pepper. Trim and clean the mushrooms. Dredge them with flour all over, shaking off excess, then dip in the beaten eggs and coat all over with the breadcrumbs. Handle the more delicate varieties of mushrooms gently as you prepare them, to avoid damaging or breaking them. Set them on a large platter or chopping board. Heat scant 1 oz (20 g) butter and 2 tbsp oil in a large, non-stick frying pan; when the butter has stopped foaming the fat and oil will be hot enough. Fry the mushrooms one at a time for about 3 minutes on each side, until the coating is crisp and golden brown. Drain on kitchen paper and keep hot, uncovered, in the oven while you fry the rest. Add the same quantity of butter and oil to the pan, waiting for it to reach the correct heat before frying the next mushroom. Serve without delay, with fresh lemon wedges and parsley.

———•———

MUSHROOM RISSOLES WITH FRESH TOMATO SAUCE

1–1¼ lb (500 g) chanterelles, field mushrooms or cultivated mushrooms

1 sprig thyme

1 clove garlic

2 whole eggs

1½ oz (40 g) grated Parmesan or other hard cheese

4 tbsp fine fresh breadcrumbs

2 tbsp finely chopped parsley

light olive oil

salt and freshly ground pepper

Garnish with:

sprigs of parsley

Serve with:

fresh tomato sauce (see page 258)

Make the sauce; set aside while you prepare the rissoles. Trim and wipe or rinse and dry the mushrooms; cut them in fairly large pieces and place in a large, non-stick frying pan with 2 tbsp oil, the thyme and the whole peeled garlic, slightly crushed with the flat of a heavy knife. Sprinkle with a little salt and pepper, cover tightly and cook over a very low heat for about 20 minutes or until the mushrooms are very tender. Remove the lid and cook over a high heat so that all the moisture evaporates. Remove and discard the thyme and garlic. Chop the mushrooms finely.

Drain off any remaining liquid from the mushrooms and mix them in a bowl with the lightly beaten eggs, a little more salt and pepper, the grated cheese, breadcrumbs, a very little crushed garlic if wished and the parsley. Blend into a dense, even consistency that will hold its shape, adding more breadcrumbs if necessary. Divide the mixture into even portions and shape these into round, flattened rissoles looking rather like fishcakes, about 2 in (5 cm) in diameter. Shallow-fry, turning once. Drain well and keep hot on kitchen paper to absorb more oil while you finish frying the rest. Serve garnished with parsley sprigs and accompanied by the fresh tomato sauce.

———•———

TRUFFLED EGGS WITH POLENTA

1 small fresh or tinned truffle or tinned truffle pieces or peelings

8 very fresh eggs

1 oz (30 g) butter

1 quantity polenta (see page 262)

salt and pepper

Make the polenta and keep hot. Divide the butter in half and melt one piece in each of two small non-stick frying pans or fireproof eared dishes over a very low heat or heat diffusers. Break 4 eggs carefully into each pan or dish so that they sit neatly beside one another. Cook gently for about 2–3 minutes (longer if you are using a heat diffuser). Season

263

the whites only with a little salt and pepper when they have just set and turned opaque. Traditionally the yolks should still be liquid; if you want to cook them for longer so that they start to set, cover with a lid.

Spoon the piping hot polenta on to heated plates; carefully slide two eggs on to each polenta portion. Sprinkle with the truffle, sliced wafer-thin. If you are lucky enough to have a fresh truffle add extra flavour by leaving the eggs (in their shells) with the whole cleaned truffle in an airtight container in the refrigerator or a cool larder for up to 24 hours before cooking this dish. The truffle scent will permeate the eggs through their shells.

———— • ————

MUSHROOMS BORDEAUX STYLE

10 oz (300 g) cultivated mushrooms
2 tbsp light olive oil
1 oz (30 g) butter
3 shallots, very finely chopped
3 tbsp fine fresh breadcrumbs
2 tbsp chopped parsley
1 tbsp chopped chives
salt and freshly ground pepper
Serve with:
Rösti Potatoes (see page 195) or herb omelettes

Trim and wipe the mushrooms with a damp cloth or rinse and dry thoroughly with kitchen paper. Cut each mushroom in quarters. Heat the oil and butter in a large frying pan or shallow saucepan until the butter has stopped foaming. Add the mushrooms and sauté over a moderate heat for 3–4 minutes. Season with salt and pepper. Add the shallot, followed by the breadcrumbs, and continue frying for a further 2–3 minutes.

Draw aside from the heat, stir in the parsley and chives and serve at once on heated plates.

MOREL MUSHROOMS AND ASPARAGUS WITH MOUSSELINE SAUCE

24 fresh or dried soaked morel mushrooms (morchella elata or morchella esculenta)
16 large asparagus spears
scant 1 oz (20 g) butter
3 tbsp dry white wine
7 fl oz (200 ml) mousseline sauce (see page 107)
salt
Cauliflower Timbale (see page 83) or poached fillets of white fish

If you feel uncertain about gathering your own morels (the edible type is readily distinguishable from the false morel which should not be eaten), simply buy dried ones and soak them in warm water for 20–30 minutes before using. Swish them about in the soaking water and rinse briefly to make sure that every grain of grit has been extracted from their pocketed heads. Fresh morels have more flavour and should be brushed meticulously with a soft, dry brush, not washed. This recipe can also be prepared with ordinary field mushrooms or fresh cultivated mushrooms.

Prepare the asparagus (see page 17) and boil in salted water for 12 minutes, or steam, until tender. Drain and keep hot. Trim and prepare the mushrooms. Separate the stalks from the heads or caps, cut the stalks into thick sections and cut the caps into quarters. Cook gently in the butter for about 8 minutes, sprinkling with a little salt and pepper. Sprinkle the wine over them and continue cooking for a minute or two, or until they are tender. Keep hot while making the mousseline sauce. Work quickly once this is ready: arrange the asparagus spears on heated plates so that 2 pairs of spears bisect one another, distribute the morels in the 4 empty sections left on each plate by this arrangement and spoon in a little sauce. Besides being a good vegetable accompaniment, this dish also makes a very good starter served on its own.

·Menus for Entertaining·

WINTER SUPPER

RUSSIAN NEW YEAR'S EVE DINNER

RUSSIAN CARROT AND GREEN
APPLE SALAD

•

CARROT BREAD (BABKA)

•

BAKED POTATOES WITH
CAVIAR AND SOURED CREAM

•

RUSSIAN BEETROOT AND
CABBAGE SOUP (BORSCHT)

•

JERUSALEM ARTICHOKE VOL-
AU-VENTS RUSSIAN-STYLE

•

Chilled vodka or dry white wine

Champagne

Black Russian

New Year's Day Brunch

Celery and Green Apple
Juice

•

Popovers

•

Sweet Potato Pancakes
with Horseradish and
Apple Sauce

•

Mung Dhal Fritters

•

Pumpkin and Amaretti
Cake

•

English breakfast tea

Colombian coffee

Pomegranate Fizz

AFTERNOON TEA

HAM AND CRESS SANDWICHES

•

VEGETABLE SAMOSAS

•

RHUBARB CAKE

•

CARROT CAKE

•

CHESTNUT CAKE

•

ITALIAN EASTER PIE

•

Spiced Indian tea

Orange Pekoe, Earl Grey,
Lapsang Souchong

279

*Until fairly recently afternoon tea
had fallen a little out of favour and
become a half forgotten meal but it
is now starting to stage a revival. Whether it is
tea on the lawn on a lazy summer afternoon or a
more substantial high tea served at about 6
o'clock, the evening equivalent of brunch with
everyone sitting round the table, a tea party
provides the perfect occasion for a get-together.
Everything can be prepared well in advance;
only the tea must be made at the last minute and
a fresh pot brewed when necessary. You can
entertain as many or as few people as you like,
inviting just two or three friends, or plan a large
tea party, or even a garden party at which
children will be more than welcome.
Included in the menu will be a choice of teas:
Orange Pekoe, Earl Grey or Lapsang Souchong,
and traditional items such as cucumber
sandwiches; the rhubarb cake will be a popular
choice, agreeably moist and not too sweet, while
the chestnut cake will be an unfamiliar and
unusual treat for most of your guests. The
children might for once be persuaded to forgo
their favourite fizzy drinks in favour of healthier,
more nourishing alternatives such as homemade milk shakes or fruit-flavoured teas, served
with ham and cress sandwiches and carrot cake.
For a high tea, or for those who are very hungry
and enjoy a savoury dish at any time of day,
Italian Easter Pie, accompanied by a cup of
Queen Mary's tea with its intriguing muscatel
flavour will be a great treat. Vegetable samosas,
crisp pastry parcels enclosing a spicy pea and
potato filling, or sandwiches with a filling of
mildly curried chopped egg mayonnaise served
with a glass of spiced Indian tea would add an
exotic touch.*

CARROT CAKE	p. 199
ITALIAN EASTER PIE	p. 50
HAM AND CRESS SANDWICHES	p. 49
VEGETABLE SAMOSAS	p. 210
RHUBARB CAKE	p. 96

CHESTNUT CAKE

Preparation: 15 minutes

Cooking: 30 minutes

Easy

9 oz (250 g) chestnut flour

18 fl oz (500 ml) milk

3½ oz (100 g) caster sugar

scant 1 oz (20 g) pine nuts

1 tbsp fresh rosemary leaves or generous pinch powdered rosemary

3½ fl oz (100 ml) light olive oil

pinch salt

Yields 16 portions. Preheat the oven to 400°F (200°C) mark 6. Brush the inside of 2 shallow sandwich cake tins about 10½ in (27 cm) in diameter lightly with extra olive oil. Sift the chestnut flour into a large mixing bowl and gradually stir in the milk,

mixed with an equal volume of cold water. Add a large pinch of salt and the sugar, stir well and then beat in the olive oil gradually with a balloon whisk, adding a little at a time. Spread out half the mixture in each prepared cake tin, smoothing it level with a palette knife. Sprinkle the pine nuts, rosemary leaves and a very little olive oil evenly over the surface.

Bake in the oven for about 30 minutes or until the surface has turned a good golden brown and a skewer or knife pushed into the cake comes out clean and dry. Leave to cool for about 10 minutes, then loosen by running the palette knife round the inside of the tins and turn out.

———•———

SPICED INDIAN TEA

Preparation: 5 minutes

and 15 minutes' simmering

10 cardamoms

2 cinnamon sticks

10 cloves

2½ tbsp sugar

10 fl oz (300 ml) milk

3 tbsp Ceylon or Assam tea

Make a small slit in the tough skin of the cardamoms so that they release their flavour but not the seeds enclosed inside. Place the cardamoms, cinnamon sticks and cloves in a lipped saucepan, add 2½ pints (1.4 litres) cold water, bring to the boil, cover and simmer gently for 10 minutes. Add the sugar and milk and return to the boil; remove from the heat and immediately add to the tea in a warmed teapot. Leave to stand for 2 minutes and then serve at once through a fine strainer. Serves 6.

Chestnut cake

MARDI-GRAS BUFFET SUPPER

SUN-DRIED TOMATOES IN OIL
WITH BREAD AND CHEESE

•

CRUDITÉS WITH SESAME DIP

•

MUSHROOM AND TRUFFLE
SALAD WITH SHERRY
MAYONNAISE

•

CHICKEN AND CHICORY
GALETTE

•

PASTA CAPRI

•

MUSHROOM AND SPINACH
STRUDEL WITH RED WINE
SAUCE

•

CAULIFLOWER TIMBALE

•

GLOBE ARTICHOKES ROMAN
STYLE

•

STRAWBERRY BAVAROIS WITH
KIWI FRUIT SAUCE

•

NOUGAT PASTRIES

ROMANTIC SUPPER FOR TWO

SESAME STRAWS

———•———

AIDA SALAD

———•———

CURRIED CREAM OF
CAULIFLOWER SOUP

———•———

AUBERGINE TIMBALES WITH
FRESH MINT SAUCE

———•———

MACEDOINE OF VEGETABLES
WITH THYME

———•———

Chris Evert

Stinger

INDIAN DINNER

DEEP-FRIED OKRA

•

INDIAN SPICED CABBAGE AND
POTATOES

•

AUBERGINE AND MINT RAITA

•

PEAS AND FRESH CURD
CHEESE INDIAN STYLE

•

RED LENTILS WITH RICE
INDIAN STYLE

•

INDIAN PEA SOUP

•

SAFFRON MILK JELLY WITH
ROSE PETALS

•

Bombay

Mango Fizz

Darjeeling tea

PROVENÇAL SUMMER LUNCH

SALADE GOURMANDE WITH
SHERRY VINAIGRETTE

·

CHILLED TOMATO SOUP WITH
SCAMPI

·

PROVENÇAL STUFFED
VEGETABLES

·

APPLE SORBET WITH
CALVADOS

·

CHOCOLATE DATE BARS

ORIENTAL FEAST

TURKISH STUFFED VINE
LEAVES

•

COURGETTE FRITTERS

•

TSATSIKI

•

CARROT SALAD TURKISH
STYLE

•

TURKISH AUBERGINE DIP

•

TURKISH STUFFED PEPPERS

•

ALMOND CREAM WITH
POMEGRANATE SEEDS

•

Raki

Iced vodka

Southern Anatolian coffee

GREEK SUMMER LUNCH

GREEK SALAD

•

GREEK CHICK PEA
CROQUETTES

•

DEEP-FRIED GREEK
AUBERGINE TURNOVERS

•

GREEK SPINACH AND FETA
CHEESE PIE

•

STUFFED AUBERGINES GREEK
STYLE

•

TSATSIKI

•

LEMON AND BASIL SORBET

•

RISÓGALO

•

Japanese Dinner Party

*N*ow that Japanese food is becoming increasingly popular, many Western cooks are starting to appreciate this intricate and artistic cuisine which also offers a very healthy diet. Japanese cooking techniques lend themselves most readily to the painstaking cook who likes to devote a lot of time and attention to food preparation: the results can be very rewarding, taste delicious and look very beautiful.

Welcome your guests with a Mikado cocktail. The starter of daikon root and salmon roe will look and taste beautiful; Japanese rice wine or saké, gently warmed and served in tiny cups, can be served with this, switching to chilled beer or green tea to accompany the rest of the meal. For the next course, a salad of mushroom and soy bean flour noodles, makes a relatively bland prelude to the tempura, fresh tender morsels of vegetables dipped in featherlight batter and deep fried; served with tentsuyu sauce; you can leave your guests to chat and drink more saké, beer or green tea while you fry these. If you prefer not to set about cooking while your guests wait, you could omit the tempura and go straight on to the soy bean sprouts with pork and steamed rice. The main part of the meal will end with miso and tofu soup. After a pause, it is time for the dessert course: a refreshing melon sorbet. Round off the meal with small diamond- or triangle-shaped slices of aduki bean jelly (recipe on page 240) accompanied by tiny glasses or porcelain cups of Japanese plum brandy.

1   2   2   3   1   4   5

MELON SORBET
Misore no moto

Preparation: 15 minutes + 3 hours for freezing

Easy

12 oz (350 g) peeled melon flesh

9 oz (250 g) caster sugar

Decorate with:

small mint leaves

Purée the melon flesh in the blender or food processor with 12 fl oz (350 ml) cold water and the sugar for 30 seconds at high speed, until smooth. If you have an electric ice-cream maker, pour the melon purée into the container and process for 15 minutes after which the sorbet will be frozen; transfer the container to the freezer. If you do not have an ice-cream maker, pour the purée into a freezerproof container with a tight-fitting lid and place in the freezer in the fast freeze compartment if you have one; take the bowl out after about 15 minutes and beat for about 1 minute with a hand-held electric beater; replace in the freezer for another 15 minutes and then repeat the operation. By now the sorbet mixture should be thick. Repeat twice more. When it is time to serve the sorbet, scoop out 3 balls for each serving with a small ice cream scoop and place in tall glasses, small dishes or coupes; decorate with mint.

MIKADO

3 fl oz (80 ml) brandy

4 dashes lemon barley water

4 dashes Crème de Noyau

2 dashes Angostura bitters

4 dashes Curaçao

Place the ingredients in the shaker with ice and pour into chilled cocktail glasses. Serves 2.

Melon sorbet

SOUTH AMERICAN THÉ DANSANT

311

A party with a South American theme is guaranteed to cheer everyone up. It could also be an imaginative way of celebrating Hallowe'en or Thanksgiving. The menu on these two pages is intended for sixteen people, while the recipes which come from other sections can simply have their ingredients multiplied by four. With the exception of the rice, all the dishes can be prepared well in advance and if you choose to have a buffet, the easiest solution, you can serve all the savoury dishes at once. Your guests will probably want to start with the guacamole dip with tortilla chips; then comes the iced pumpkin soup which can be attractively served in a hollowed out half pumpkin. Stuffed sweet peppers topped with soured cream, coriander and pomegranate seeds come next, and a substantial and filling Creole dish of beans with rice and fried bananas. The Caipirinha cocktail should make everyone sufficiently uninhibited to dance or at least sway to the most carefree South American music while those who prefer a long, thirst-quenching drink can enjoy a Cuba libre. Chilled lager or some good Chilean wine would be other options.

1   2   3   2   4

MEXICAN PEPPERS WITH POMEGRANATE SAUCE	p. 138
BEANS CREOLE WITH RICE AND FRIED BANANAS	p. 220
CHILLED ICED PUMPKIN SOUP	p. 118
GUACAMOLE	p. 113

FROZEN PEACH MOUSSE WITH CARAMEL SAUCE

Preparation: 25 minutes + 3 hours for freezing
Cooking: 20 minutes
Easy
2¼ lb (1 kg) caster sugar
2¼ lb (1 kg) ripe peach flesh, fresh or tinned
3½ pints (2 litres) whipping cream
5 oz (150 g) crumbled ratafias or macaroons

For the caramel sauce:

10 oz (300 g) caster sugar

almond essence

Place the sugar in a heavy-bottomed saucepan with 1 pint (600 ml) cold water and heat slowly to boiling, stirring occasionally. Simmer for 1 minute, then take off the heat and leave to cool. Measure out exactly 1 pint 16 fl oz (450 ml) of this sugar syrup, saving any surplus for other recipes such as fruit salad. Leave the syrup to cool at room temperature in a very large mixing bowl. Peel the peaches, remove the stones and measure out the required amount of flesh. If using tinned peaches, drain them very thoroughly beforehand. Place the peach flesh in the food processor or blender and process to a smooth purée. Put this purée through a fine sieve and measure out exactly 1 pint 16 fl oz (450 ml) of it. When the syrup is completely cold, stir in the peach purée. Beat the well chilled cream until stiff and fold into the peach and syrup mixture together with the fairly finely ground amaretti. Pour into rectangular containers, cover with greaseproof paper and freeze for at least 3 hours before turning out on to a large, chilled serving platter and cutting into slices. Serve without delay.

While the mousse is freezing, make the sauce: place the sugar in a heavy-bottomed saucepan with 7 fl oz (200 ml); bring slowly to the boil, stirring now and then with a wooden spoon. Simmer over a low heat, stirring frequently, for about 20 minutes or until the syrup has started to caramelize and has turned a light golden brown; draw aside from the heat; add 7 fl oz (200 ml) cold water, taking care to keep your hands and face out of the way of the sudden burst of hot steam; add 4 drops almond essence. Leave to cool to room temperature. Serves 24.

CAIPIRINHA

2 fl oz (50 ml) light rum

1 small lime

2 tsp cane sugar

Cut the lime in quarters and pound these in a mortar with the sugar using a pestle. Discard the skin, save the juice and sugar and place in a glass with ice cubes and the rum, mix and serve.

———•———

CUBA LIBRE

2 fl oz (60 ml) light rum

juice of ½ lime

chilled Coca-Cola or Pepsi-Cola

Pour the rum into a tall glass, add the lime juice and top with Coca Cola or Pepsi Cola. Serves 1.

Frozen peach mousse with caramel sauce

·Index·